Study Guide for Moore's

THE BASIC PRACTICE OF STATISTICS

William Notz
Ohio State University

Rebecca L. Busam

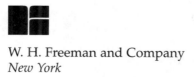

W. H. Freeman and Company
New York

ISBN: 0-7167-2682-3

Printed in the United States of America

1 2 3 4 5 6 7 8 9 0   VB   9 9 8 7 6 5

CONTENTS

INTRODUCTION

Statistics and this Study Guide

Who is the greatest baseball player of all time? Do cellular phones cause brain damage? Are we experiencing global warming? Do some people have ESP? Should handgun sales be regulated? Each day scientists, legislators, sports fans, and many other types of people attempt to answer questions like these. The strongest and most convincing arguments are the ones backed up by data and not just opinion. How well do the data support the argument? The study of statistics will give us the answer.

Understanding the basic concepts of statistical reasoning will help us assess the quality of any argument based on data. It is our goal with this *Study Guide* for *The Basic Practice of Statistics*, to aid in teaching you these concepts and to give you practice reasoning with data. This can be an enjoyable and rewarding experience, especially when examining data that are interesting to you personally. Since governments often make policy decisions and pass laws based on arguments from data, a knowledge of how to reason properly from data is also crucial for you as a citizen affected by these decisions. It is amazing when you consider how often important decisions are made using faulty statistical reasoning.

Features of this Study Guide

A more immediate purpose of this *Study Guide* is to help you learn and review the material in the textbook. Each chapter follows a set structure: There is an overview of each section, followed by detailed solutions to selected textbook exercises. Next are solutions to key chapter review exercises. Each chapter concludes with an original case study illustrating how the material in the chapter can be applied to a real-life situation. Some case studies also show how the material can be used to critique the improper uses of statistics. A glossary of terms and symbols used in the textbook appears at the end of the *Study Guide*.

The section overviews might be particularly useful as you review coverage in preparation for tests. We have summarized the information that we believe is most important. However, keep in mind that your instructor may emphasize materials in a different way. This *Study Guide* is meant to add to the usefulness of the textbook coverage and your instructor's lectures, thereby helping you to excel in this course. In no way can it be thought of as a replacement for the textbook or your lecture notes. Obviously if anything in the overviews seems unfamiliar, you should immediately go back and re-read the textbook and your lecture notes.

It would also be a mistake to think that reading our solutions to the textbook exercises can replace working them out by yourself. This would be the same as if you read a book about swimming and then believed you were prepared to jump into the deepest end of the pool. Real learning takes work—you learn by doing. We suggest you try to solve the exercises on your own first. Then use our solutions to check your work. If you have gotten the solution wrong, use ours to help you understand why.

One way to use our solutions to key chapter review exercises is as part of a simulated test. Try putting time limits on how long you have to solve exercises. After you have come up with answers, check our solutions. If you have trouble, you need to study further. Of course, don't neglect your instructor as a resource if you have problems in the course. Face-to-face communication with your instructor is an invaluable way to clear up difficulties.

The case studies in each chapter build off the real-life emphasis of your textbook. They are meant to demonstrate how statistics can be applied to everyday life. They can also help you in learning how to evaluate the improper use of statistics. As such they can serve to extend one of the main goals of *The Basic Practice of Statistics*—to help you become a better critical thinker and consumer of statistics. The case studies do all of this while also reinforcing the topic coverage of the chapters.

Acknowledgments

The real data sets used in the case studies were collected as part of a project sponsored by the National Science Foundation. Professor Elizabeth Stasny, Professor Dennis Pearl, Professor Paul Velleman, as well as students Michael Bowcutt, David Butler, Jim Clark, Nicole DePriest, Eric Eastmo, Greg Elfring, Kathy Fritsch, Matt Hutcheson, H-C Tsai, John Walker, and Mark Zabel were all members of this project, and as such are contributors to this *Study Guide*. We acknowledge their efforts and thank them for their work. We would also like to thank David S. Moore for his encouragement and assistance during this project. His writings and teaching have served as models for us over the years.

William Notz
Columbus, Ohio

Rebecca L. Busam
Chicago, Illinois

CHAPTER 1

EXAMINING DISTRIBUTIONS

CHAPTER OVERVIEW

Data are everywhere you look. By data we mean a list of measurements of a specific characteristic of objects or individuals. This might be a list of the batting averages of all players in the current baseball season like those typically reported in the Sunday sports section of your local paper. It might be the number of crimes reported each day for a given city for a period of several years. It might be the thickness of the ozone layer over the North Pole measured annually for each year of this century. The goal in looking at any such data is to uncover important patterns and interesting features. We should envision ourselves as explorers looking for hidden treasures. Like any such explorer, not all data will have hidden gems, but occasionally one does strike gold.

Here is the basic strategy. First look at the data. Are they a list of names? Numbers? Next look at simple graphs. These will give you an overall sense of the data and will alert you to interesting patterns or features of the data. Follow this by looking at numerical summaries of data which focus on some of these specific features of the data. Some numerical summaries are the mean, median and variance. Finally, attempt to summarize what the data seem to be saying.

This exploration usually does not take place in a vacuum. Data have a context. They are usually collected to answer some question. Understanding the story behind the data and the purpose for which they were collected is important and will help guide your exploration. But don't let it completely restrict you either. Knowledge of the purpose for which the data were collected can sometimes adversely affect our judgment if we are not careful. We sometimes see patterns because we want to see them.

SECTION 1.1

SECTION OVERVIEW

Chapter 1 of *Basic Practice of Statistics* discusses several methods for exploring data after looking over the data and digesting the story behind the data. The first methods are simple graphs which give an over all sense of the pattern of the data. What graphs are appropriate depend on whether the data are numerical or not. Non-numerical data use **bar charts** or **pie charts**. Numerical data use **histograms** or **stemplots**. Numerical data collected over time use a **timeplot** in addition to a histogram or stemplot. When examining the data through graphs we should be on the alert for

- unusual values that do not follow the pattern of the rest of the data,
- some sense of a central or typical value of the data,
- some sense of how spread out or variable the data are,
- some sense of the shape of the overall pattern.

In timeplots, be on the lookout for **trends** over time. These features are important whether we draw the graphs ourselves or they are done for us by a computer.

KEY CONCEPTS

Types of variables

There are several different types of variables you need to remember. Each type requires different handling.

- **categorical variable**—a variable that records to what group or category an individual belongs.

  ➤ Hair color is a categorical variable.

- **quantitative variable**—a variable which has numerical values and with which it makes sense to do arithmetic. It is a quantity.

  ➤ The number of magazines subscribed to is quantitative. We could ask for an average number of subscriptions or a maximum.

How to draw a histogram

1. Divide the range of values of the data into classes or intervals of equal length.

   ➤ If you ask 40 people to pick a number from one to ten, the data could be put into intervals in a variety of ways. One possibility is the following.

   1 - 2
   3 - 4
   5 - 6
   7 - 8
   9 - 10

2. Count the number of values of our data that fall into each interval.

➤ The responses of our 40 people might look like the following.

Interval	Count
1-2	3
3-4	10
5-6	8
7-8	12
9-10	7

3. Draw the histogram.

a. Mark the intervals on the horizontal axis and label the axis. Include the units.

b. Mark the scale for the counts on the vertical axis. Label the axis.

c. Draw bars, centered over each interval, up to the height equal to the count. There should be no space between the bars (unless the count for a class is zero, so that its bar has height zero).

➤ Here is the finished histogram.

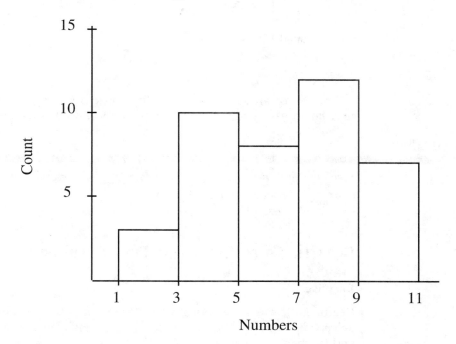

Once you have finished creating a histogram there are four things to look for:

• **outliers**—observations that stand out from the rest for some reason

• **center**—the "middle" of the data

• **spread**—the range; the extent of the data; how far the values are from each other

• **shape**—what the distribution looks like

Drawing a stemplot

1. Put the observations in numerical order.

2. Decide how the stems will be shown. Commonly, a stem is all digits except the rightmost. The leaf is then the rightmost digit.

3. Write the stems vertically in increasing order. Write each stem only once. Draw a vertical line next to the stems.

4. Write each leaf next to its stem.

5. Rewrite the stems and put the leaves in increasing order.

➣ Suppose our data are the time, in minutes, it took to complete a homework assignment. The data are

| 45 | 44 | 72 | 86 | 51 | 50 | 43 | 65 | 72 | 80 |
| 31 | 60 | 67 | 56 | 71 | 75 | 41 | 39 | 40 | 49 |

The stemplot looks like

```
30 | 1 9
40 | 0 1 3 4 5 9
50 | 0 1 6
60 | 0 5 7
70 | 1 2 2 5
80 | 0 6
```

The stems can be rounded or split in order to change the number of stems, which may be necessary to highlight any patterns and unusual observations.

REMEMBER:

Stemplots are most useful for smaller data sets. For larger data sets stemplots can be difficult to draw and histograms are preferred.

SOLUTIONS TO SELECTED TEXT EXERCISES

Exercise 1.1

Region is clearly a categorical variable since it is not numerical. The remaining variables (population (1000), SAT verbal, SAT math, percent taking, dollars per pupil, and teacher's pay) are all quantitative.

Exercise 1.3

(a) The overall shape is symmetric. While the histogram is not perfectly symmetric, this is quite symmetric for real data.

(b) Since the histogram is symmetric, we would place the center at the bar covering the 10 to 20% total return. A slightly more careful inspection indicates that this central bar includes about the middle 21% of all NYSE stocks and so includes the center. Technically, this is as precise as we can be about the value of the center. If pressed for a more specific value, we would be tempted to say that a total return of about 15% is the center. This assumes, however, that the total return on stocks in the middle bar is roughly evenly distributed between 10 and 20%. We don't know this, although it seems like a reasonable assumption.

(c) The smallest return is between −70 and −60%. The largest return between 100 and 110%. Once again, if we can assume returns on stocks in these ranges are evenly distributed, we could say the smallest return is approximately −70% and the largest 110%.

(d) Adding the heights of the bars to the left of 0, we would estimate that about 23% or 24% of all stocks lost money. This answer is approximate since it is difficult to estimate bar heights exactly (they need not be an integer value).

Exercise 1.7

Here is a stemplot of the data using the first two digits (the hundreds and tens place) as the stems.

```
10| 1 3 9
11| 5
12| 6 6 9
13| 7 7
14| 0 8
15| 2 4 4
16| 5 5
17| 8
18|
19|
20| 0
```

There is an outlier at 200. The center of the distribution between 137 and 140, since half (9) of the observations are 137 or less, and half are 140 or more. If we ignore the outlier at 200, the data range from the smallest value of 101 to the largest value of 178, for a range of about 75–80.

Exercise 1.9

(a) Here is a time plot of the data.

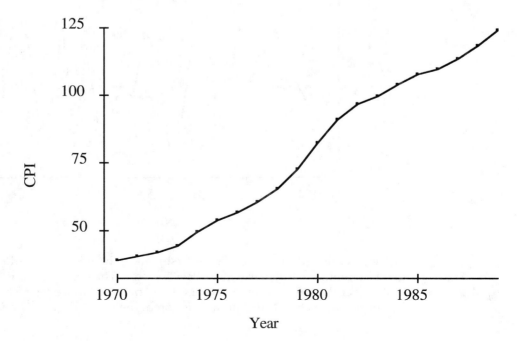

(b) The overall trend is upwards. There are no years in which this trend was reversed.

(c) Prices were rising fastest in the middle years, 1978–1982 (especially from 1979–1980). They appear to be rising most slowly in the early portion of the period, 1970–1973.

Exercise 1.15

(a) If we use the units and tenths digits as stems we will have about 35 stems (1.1 up to 4.6). This is too many stems. We reduce this number by rounding to the nearest tenth. This produces only four stems, which seems like too few. We therefore split these to obtain eight stems. The resulting stemplot is

```
1| 1 2
1| 5 5 6 7 9 9
2| 0 1 1 1 1 2 2 2 3 3
2| 5 6 8 8 9 9
3| 0 1 1 3
3| 5
4|
4| 7
```

The general shape is symmetric. There is an outlier at 4.7 inches. Ignoring the outlier, the plot is centered around 2.2 inches and has a spread of about 2.4 inches, (the smallest value being about 1.1 inches and the largest 3.5 inches).

(b) Here is a time plot of the data.

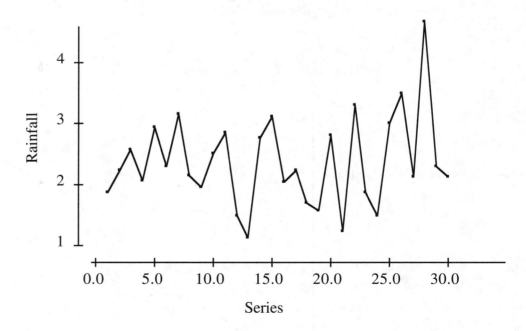

The overall pattern fluctuates. At first glance, there is a suggestion of an upward trend in rainfall, particularly in the last 10 years. The peaks of the fluctuations seem a bit higher than in the previous years. This "pattern," however, rests strongly on the outlier of 4.69 inches. If this point is ignored, the upward trend is not as clear.

SECTION 1.2

SECTION OVERVIEW

Once you examine graphs to get an overall sense of the data, it is helpful to look at numerical summaries of features of the data, which make more precise the notions of center and spread.

Measures of center: • **mean**
                             • **median**

Measures of spread: • **quartiles**
                             • **standard deviation**
                             • **variance**

For measures of spread, the quartiles are appropriate when the median is used as a measure of center. In fact, the **five-number summary**, reporting the largest and smallest values of the data, the quartiles and the median, provides a nice, compact description of the data. The five-number summary can be represented graphically by a boxplot. If you use the mean as a measure of center, then the standard deviation and variance are the appropriate measures of spread. Note that means and variances can be strongly affected by outliers and are harder to interpret for skewed data. The median and quartiles are more appropriate when outliers are present or when the data are skewed.

KEY CONCEPTS

Finding the mean, $\bar{x}$

The mean is the common arithmetic average. If there are n observations, $x_1, x_2, \ldots, x_n,$ then the mean is

$$\bar{x} = \frac{x_1 + x_2 + \ldots + x_n}{n} = \frac{1}{n} \sum x_i$$

Recall that $\sum$ means "add up all these numbers."

Finding the median

1. List all the observations from smallest to largest.

2. If the number of observations is odd, then the median is the middle observation. Count from the bottom of the list of ordered values up to the $(n + 1)/2$ largest observation. This observation is the median.

3. If the number of observations is even, then the median is the average of the two center observations.

➢ Using the homework time data (see Drawing a stemplot) we find the median (in bold) is 56. The ordered data are:

31 39 41 43 44 45 49 50 51 **56** 60 65 67 71 72 75 80 86

n = 19 so we selected the $(n + 1)/2 = $ 10th largest observation, 56.

Finding the quartiles, Q_1 and Q_3

1. Locate the median.

2. The first quartile, Q_1, is the median of the lower half of the list of ordered observations.

3. The third quartile, Q_3, is the median of the upper half of the list of ordered values.

➤ Again, for the homework time data (median in bold, quartiles underlined)

31 39 41 43 <u>44</u> 45 49 50 51 **56** 60 65 67 71 <u>72</u> 75 80 86

so $Q_1 = 44$ and $Q_3 = 72$.

Finding the variance, s^2 and standard deviation, s

1. Take the average of the squared deviations of each observation from the mean. In symbols, if we have n observations, $x_1, x_2, \ldots, x_n$, with mean $\bar{x}$,

$$s^2 = \frac{(x_1 - \bar{x})^2 + (x_2 - \bar{x})^2 + \ldots + (x_n - \bar{x})^2}{n - 1} = \frac{1}{n - 1} \sum (x_i - \bar{x})^2$$

(Remember, Σ means "add up")

If you are doing this calculation by hand, it is best to take it one step at a time. First calculate the deviations, then square them, next sum them up, and finally divide the result by $n - 1$.

2. The standard deviation is the square root of the variance, i.e., $s = \sqrt{s^2}$. Some things to remember about the standard deviation.

- s measures the spread around the mean,
- s should only be used with the mean, not the median,
- if $s = 0$, then all the observations must be equal,
- the larger s is, the more spread out the data are,
- s can be strongly influenced by outliers and it is best to use s and the mean only if the distribution is symmetric or nearly symmetric.

REMEMBER:

1. The five-number summary is best for nonsymmetric data. The median and quartiles are not influenced by outliers.

2. The mean and standard deviation are most appropriate to use only if the data are symmetric because both of these measures are easily influenced by outliers.

3. If the distribution of the data is symmetric, then the mean and median will be the same.

SOLUTIONS TO SELECTED TEXT EXERCISES

Exercise 1.21

(a) The total of the 18 scores is 2539 and so the mean is $\bar{x}$ = 2539/18 = 141.056. Verify that your calculator (or software) gives the same answer (we have rounded to three decimal places).

(b) If we remove the outlier of 200, the total becomes 2339 and the mean is $\bar{x}$ = 2339/17 = 137.588. The mean drops by about 3.5 when the outlier is removed.

Exercise 1.25

The median is $490,000 and the mean is $1,160,000. The distribution of baseball salaries is right skewed. Only a few players on each team command high salaries of several million dollars per year. Most players make salaries of a few hundred thousand dollars, closer to the minimum salary of just over $100,000. The few extremely high salaries pull up the mean.

Exercise 1.27

(a) The histogram in Figure 1.10 (page 129) is nearly symmetric but may be very slightly skewed to the right. As a result we would expect the mean to be about the same as the median.

(b) Here are the ages sorted in increasing order.

42	52	57
43	52	57
46	54	57
46	54	58
47	54	60
48	54	61
49	55	61
49	55	61
50	55	61
51	55	64
51	56	64
51	56	65
51	56	68
51	57	69

From this, it is relatively easy to calculate the five-number summary. We find

Minimum = 42
1st quartile = 51
Median = 55
3rd quartile = 58
Maximum = 69

With a calculator we find the mean to be 54.833. Indeed, the mean and median are fairly close.

(c) The middle half of the age of new presidents should fall between the 1st and 3rd quartiles, namely between 51 and 58.

Exercise 1.29

(a) The sum of the six observations is 32.4. Thus the mean is 32.4/6 = 5.4.
(b) The deviations of the six observations from the mean are

$$5.6 - 5.4 = 0.2$$
$$5.2 - 5.4 = -0.2$$
$$4.6 - 5.4 = -0.8$$
$$4.9 - 5.4 = -0.5$$
$$5.7 - 5.4 = 0.3$$
$$6.4 - 5.4 = 1.0$$

The sum of the squares of these deviations is 2.06. Thus the variance is 2.06/5 = 0.412. This has square root 0.6419, which is the standard deviation.

Exercise 1.35

A variety of graphs are possible, such as a stemplot, histogram, or boxplot. A stemplot or a histogram is likely to be the most informative. Here are both such plots, respectively:

```
4.8 | 8
4.9 |
5.0 | 7
5.1 | 0
5.2 | 6 7 9 9
5.3 | 0 4 4 6 9
5.4 | 2 4 6 7
5.5 | 0 3 5 7 8
5.6 | 1 2 3 5 8
5.7 | 5 9
5.8 | 5
```

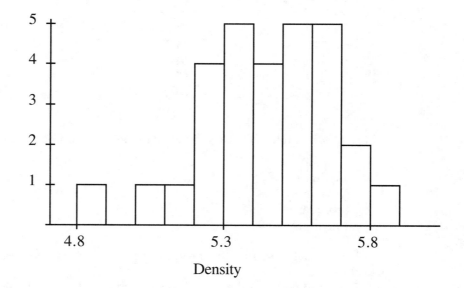

Density

The shape of the stemplot or histogram is moderately symmetric, slightly bell shaped, with no extreme outliers. It is therefore reasonable to use the mean and standard deviation to describe it. We calculate the mean and standard deviation to be

$$\bar{x} = 5.448$$
$$s = 0.221$$

The mean of 5.448 seems like a reasonable estimate of the density of the earth based on these measurements.

Exercise 1.37

A histogram, stemplot, or boxplot are possible graphs. With the diversity of values, a stemplot is challenging to make (rounding would be necessary). We settle for a histogram as perhaps the most informative graph:

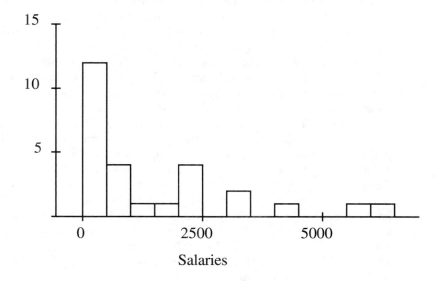

This is strongly skewed to the right, which is not unusual for salary data. The skewness suggests that the five-number summary is the appropriate numerical summary. We find

Minimum = 109
1st quartile = 158
Median = 635
3rd quartile = 2300
Maximum = 6200

One striking feature is the long tail to the right due to the few extremely large salaries. Another striking feature is the fact that five players tied for lowest salary. This is due to the fact that 109,000 was the minimum salary allowed for major league players in 1993.

SECTION 1.3

SECTION OVERVIEW

The concluding section of Chapter 1 focuses on the notion of the shape of a histogram because the histogram is a graphical way to describe the distribution of data. The shape of a histogram can be represented by smooth **density curves**. Density curves tell a great deal about the data they describe. Density curves are usually scaled so the total area under the curve is 1. Areas under density curves are then interpreted as the proportion of the data falling in some region. The balance point of the density curve (viewed as a kind of see-saw) is at the mean. The point that divides the curve into two pieces of equal area is the median. Thus the mean and median are easily estimated from the density curve. The most common density curve encountered in practice is the **normal curve**. It has a symmetric, bell shape and is a good approximation of the histograms of a surprisingly wide variety of data sets. When the normal distribution is a good description of the shape of a histogram, it can be used to answer a variety of questions about the data. In particular, we can estimate the proportion of the data falling in any range of values without having to look at the actual data! This is accomplished through the use of **standardized values**, or *z-scores*, and a **standard normal table**. This is very useful for large data sets where it would be difficult to compute this proportion directly. The normal curve illustrates the power of density curves and the importance of the shape of a histogram.

The optional section of the chapter discusses methods for assessing whether a normal curve is a satisfactory description of a given data set. The closer the data set is to a normal curve, the more reliable are statements about the data based on the normal curve.

The density curve and distribution are important concepts. They will form the basis for methods of statistical inference in later chapters. For now, remember these are descriptions of the overall pattern of our data, indicating the most commonly occurring values (represented by tall bars in the histogram) and rare values (represented by short bars). The shape says something about the relative locations of the more common and the more rare values of the data.

KEY CONCEPTS

68–95–99.7% rule

For any normal distribution approximately

> 68% of the data will be between $\mu - \sigma$ and $\mu + \sigma$
> 95% of the data will be between $\mu - 2\sigma$ and $\mu + 2\sigma$
> 99.7% of the data will be between $\mu - 3\sigma$ and $\mu + 3\sigma$

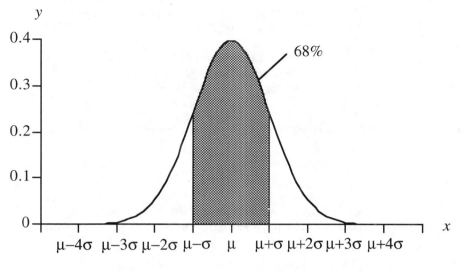

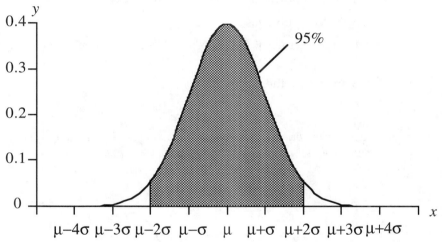

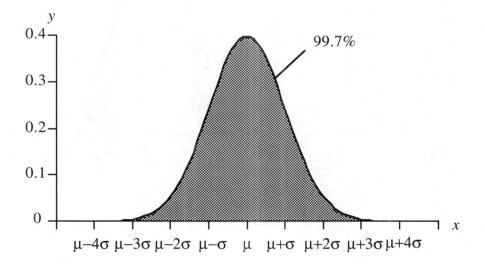

If the distribution is standard normal (remember, standard normal means $\mu = 0$ and $\sigma = 1$) then the rule is

68% of the data will be between -1 and $+1$
95% of the data will be between -2 and $+2$
99.7% f the data will be between -3 and $+3$

Standardizing or calculating the z-score

$$z = \frac{x - \mu}{\sigma}$$

where x is the observed value of the observation, μ is the mean, and σ is the standard deviation of the normal distribution the data comes from.

Finding proportions under the normal curve

1. State the problem.

2. Draw a picture of the problem. It will help you to see what area you are looking for.

3. Standardize the observations.

4. Using Table A on page 626, find the area you need.

Hint: *The normal curve has a total area of 1. The normal curve is also symmetric so areas (proportions) such as those shown below are equal.*

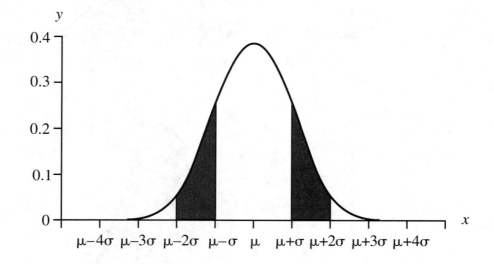

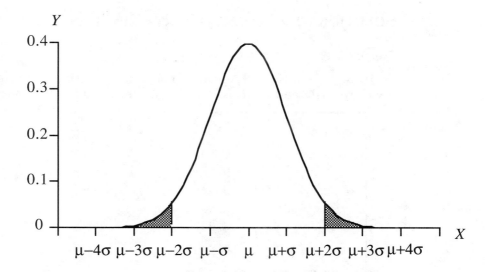

Finding a value given a proportion

1. State the problem.

2. Draw a picture to help visualize the proportions in the problem.

3. Use Table A to find the proportion closest to the one you need (look in the body of the table, not the edges.)

4. Read off the associated z-score from the table.

5. Unstandardize to get an answer in terms of the problem. Use

$$x = \mu + z\sigma$$

to unstandardize. Note that this is the reverse of how we found a proportion from a value.

REMEMBER:

You must standardize an observation in order to use the standard normal table.

SOLUTIONS TO SELECTED TEXT EXERCISES

Exercise 1.43

(a) The percent of the observations that lie above 0.8 is just the area under the density between 0.8 and 1.0. This is represented by the shaded region below.

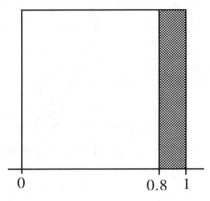

The area is easily calculated using the formula for the area of a rectangle (height × width). The area is $1 \times 0.2 = 0.2$. Therefore 20% of the observations lie above 0.8.

(b) The area of the shaded region below is the desired proportion.

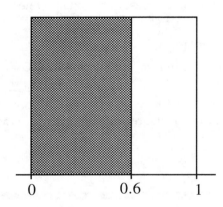

The area is $1 \times 0.6 = 0.6$. Thus 60% of the observations lie below 0.6.

(c) The area of the shaded region below is the desired proportion.

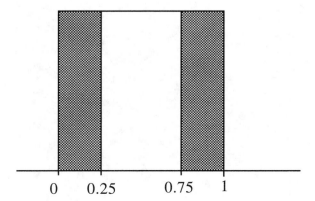

The area is $1 \times (0.75 - 0.25) = 1 \times 0.5 = 0.5$. Thus 50% of the observations lie between 0.25 and 0.75.

Exercise 1.47

(a) Since the distribution is normal, it is symmetric and the mean and median are the same. Therefore 110 is the median and hence 50% of people in this age group have scores above 110.

(b) Notice that $160 - 110 = 50$, which is twice the standard deviation of 25. Thus 160 is two standard deviations above 110. $110 - 50 = 60$ is two standard deviations below 110. The $68 - 95 - 99.7$ rule tells us that 95% of the people have scores between 60 and 160. The remaining 5% are evenly divided (since the normal distribution is symmetric) between those with scores below 60 and those with scores above 160. Half of 5% is 2.5%, therefore 2.5% have scores above 160.

(c) As indicated in part (b), the middle 95% have scores between 60 and 160.

Exercise 1.49

(a) The area under the standard normal curve that satisfies the statement $z < 2.85$ is indicated below.

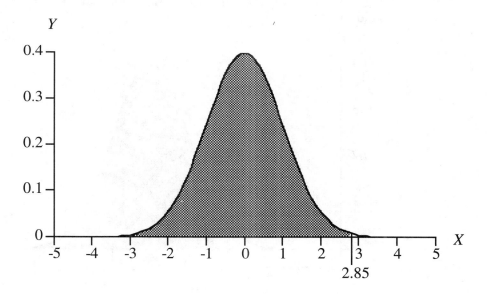

Table A gives this area as 0.9978.

(b) The area under the standard normal curve that satisfies the statement $z > 2.85$ is indicated below.

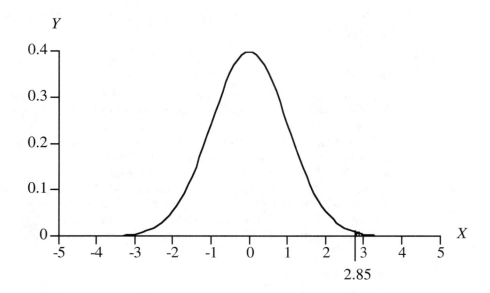

The area is $1 - $ (area to left of 2.85) $= 1 - 0.9978 = 0.0022$.

(c) The area under the standard normal curve that satisfies the statement $z > -1.66$ is indicated below.

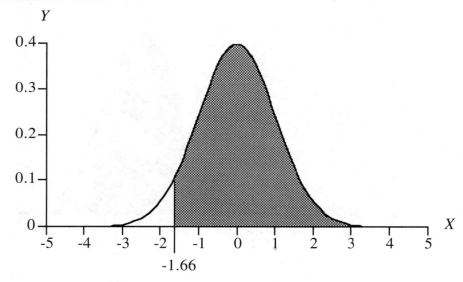

The area is $1 - $ (area to the left of -1.66). Using Table A this is $1 - 0.0485 = 0.9515$.

(d) The area under the standard normal curve that satisfies the statement $-1.66 < z < 2.85$ is indicated below.

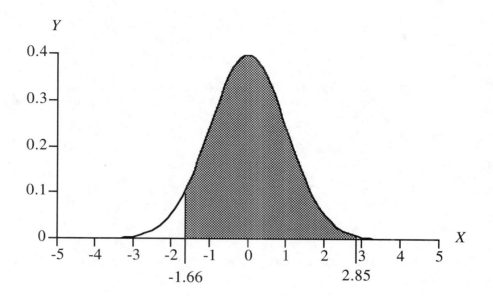

The area is (area to the left of 2.85) − (area to the left of − 1.66). Using parts (a) and (c) this is 0.9978 − 0.0485 = 0.9493.

Exercise 1.51

Follow the format suggested in the text.

(a) *State the problem.* We want the percent of men at least 6 feet (72 inches) tall, i.e., $72 \leq x$.

$$\begin{aligned} \textit{Standardize:} \quad 72 \quad &\leq \quad x \\[6pt] \frac{72-69}{2.5} \quad &\leq \quad \frac{x-69}{2.5} \\[6pt] 1.2 \quad &\leq \quad z \end{aligned}$$

This corresponds to the following area under the normal curve.

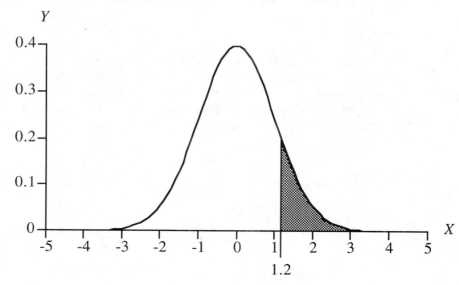

Use the table. Area above 1.2 = 1 – area below 1.2

$$= 1 - 0.8849$$
$$= 0.1151$$

Thus about 11.51% of men are at least 6 feet (72 inches) tall.

(b) *State the problem.* We want the percent of men between 60 and 72 inches tall, i.e., $60 \le x \le 72$.

Standardize:

$$60 \quad \le \quad x \quad \le \quad 72$$

$$\frac{60 - 69}{2.5} \quad \le \quad \frac{x - 69}{2.5} \quad \le \quad \frac{72 - 69}{2.5}$$

$$-3.6 \quad \le \quad z \quad \le \quad 1.2$$

This corresponds to the following area under the normal curve.

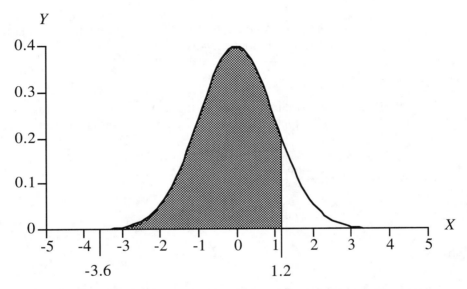

Use the table. The area between –3.6 and 1.3 = area below 1.2 – area below –3.6

$$= 0.8849 - 0 \text{ (notice } -3.6 \text{ is off the table)}$$
$$= 0.8849$$

(c) *State the problem.* We want to find the height value x with area 0.1 to its right under the normal curve with mean $m = 69$ and standard deviation $s = 2.5$. That's the same as area 0.9 to the left of x. The graph below illustrates the value we are looking for. Because Table A gives the areas to the left of z-values, we state the problem in terms of area to the left of x.

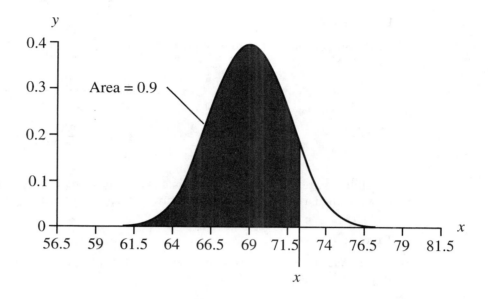

Use the table. Look in the body of the table for the entry closest to 0.9. We see it is 0.8997. This is the area below $z = 1.28$. So $z = 1.28$ is the standardized value with area 0.9 to its left.

Unstandardize. We now know that the standardized value of the unknown x is $z = 1.28$. So x itself satisfies

$$\frac{x - 69}{2.5} = 1.28$$

Solving this equation for x gives

$$x = 69 + (1.28)(2.5) = 72.2$$

An adult American male must be at least 72.2 inches tall to be among the tallest 10% of all adult American males.

Exercise 1.53

As we saw in Problem 1.35 (see solution above), the distribution is reasonably symmetric. If we calculate $\bar{x}$ and x, we find

$$\bar{x} = 5.448$$
$$x = 0.221$$

We produce the following counts

Number of observations between $5.448 - 0.221 = 5.227$ and $5.448 + 0.221 = 5.669$ is 22 ($22/29 = 75.9\%$)

Number of observations between $5.448 - 2(0.221) = 5.006$ and $5.448 + 2(0.221) = 5.890$ is 28 ($28/29 = 96.6\%$)

Number of observations between $5.448 - 3(0.221) = 4.785$ and $5.448 + 3(0.221) = 6.111$ is 29 ($29/29 = 100\%$)

The agreement with the 68–95–99.7 rule is not too bad.

Exercise 1.57

The standardized batting averages are: $\text{Cobb} = \dfrac{.420 - .266}{.0371} = 4.151$

$$\text{Williams} = \dfrac{.406 - .267}{.0326} = 4.264$$

$$\text{Brett} = \dfrac{.390 - .261}{.0371} = 4.069$$

All three are quite high. The value 4 is off our normal table, since it is unusual to observe such a large value in a standard normal distribution.

Exercise 1.61

(a) *State the problem.* We want the percent of pregnancies that last less than 240 days, i.e., $x \leq 240$.

Standardize:

$$x \quad \leq \quad 240$$

$$\dfrac{x - 266}{16} \quad \leq \quad \dfrac{240 - 266}{16}$$

$$z \quad \leq \quad -1.63$$

This corresponds to the following area under the normal curve.

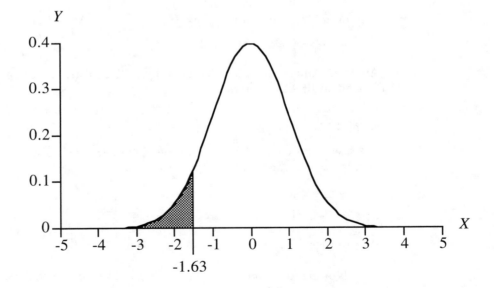

Use the table. Area below − 1.63 = .0516

Thus about 5.16% of pregnancies last less than 240 days.

(b) State the problem. We want the percent of pregnancies that last between 240 and 270 days, i.e., $240 \leq x \leq 270$.

$$\textit{Standardize:} \qquad 240 \quad \leq \quad x \quad \leq \quad 270$$

$$\frac{240 - 266}{16} \leq \frac{x - 266}{16} \leq \frac{270 - 266}{16}$$

$$-1.63 \quad \leq \quad z \quad \leq \quad 0.250$$

This corresponds to the following area under the normal curve.

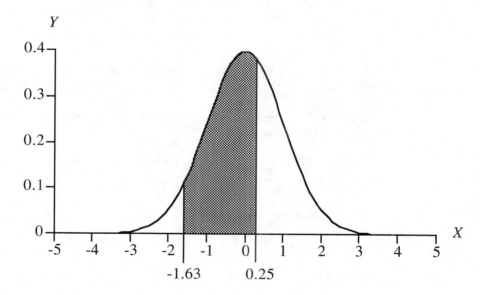

Use the table. Area between − 1.63 and 0.25 = area to left of 0.25 − area to left of − 1.63
= 0.5987 − 0.0516
= 0.5471

Thus about 54.71% of pregnancies last between 240 and 270 days.

(c) *State the problem.* We want to find the value x with area 0.2 to its right under the normal curve with mean $m = 266$ and standard deviation $s = 16$. That's the same as area 0.8 to the left of x. The graph below illustrates the value we are looking for. Because Table A gives the areas to the left of z-values, we state the problem in terms of area to the left of x.

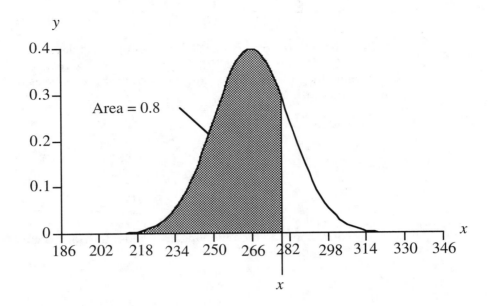

Use the table. Look in the body of the table for the entry closest to 0.8. We see it is 0.7995. This is the area below $z = 0.84$. So $z = 0.84$ is the standardized value with area 0.8 to its left.

Unstandardize. We now know that the standardized value of the unknown x is $z = 0.84$. So x itself satisfies

$$\frac{x - 266}{16} = 0.84$$

Solving this equation for x gives

$$x = 266 + (0.84)(16) = 279.4$$

A pregnancy must last at least 279.4 days to be in the longest 20% of pregnancies.

SOLUTIONS TO SELECTED TEXT REVIEW EXERCISES

Exercise 1.65

(a) Here is the stemplot (data rounded to nearest tenth of a minute and split stems).

```
06 | 8
07 | 4 4
07 | 8 8 8 8 8 9 9 9
08 | 0 1 1 2 2 3 3 3 4 4 4
08 | 5 5 5 6 6 7 7 7 7 8 8 8
09 | 0 0 0 0 1 2
09 | 8
10 | 2
```

The distribution looks roughly symmetric and there are no clear outliers.
(b) Here is a time plot of the data.

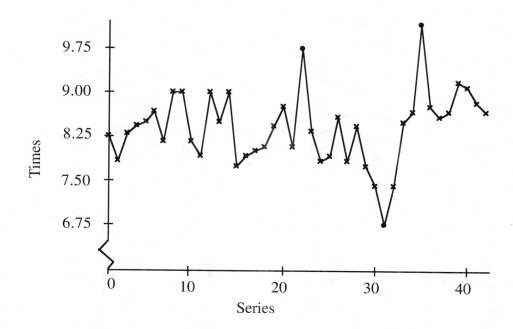

The unusually low and high times are represented by dots in the plot.

(c) Removing the three unusual observations, we find the mean and standard deviation of the remaining 39 times to be

$$\bar{x} = 8.36$$
$$s = 0.46$$

(d) We produce the following counts for all 42 points:

Number of observations between $8.36 - 0.46 = 7.90$ and $8.36 + 0.46 = 8.82$ is 25 ($25/42 = 59.5\%$)

Number of observations between $8.36 - 2(0.46) = 7.44$ and $8.3 + 2(0.46) = 9.28$ is 37 ($37/42 = 88.1\%$)

Number of observations between $8.36 - 3(0.46) = 6.98$ and $8.36 + 3(0.46) = 9.74$ is 39 ($39/42 = 92.9\%$)

The percentages are all a bit low.
If we remove the three points and recalculate we get:

Number of observations between $8.36 - 0.46 = 7.90$ and $8.36 + 0.46 = 8.82$ is 25 ($25/39 = 64.1\%$)

Number of observations between $8.36 - 2(0.46) = 7.44$ and $8.36 + 2(0.46) = 9.28$ is 37 ($37/39 = 94.9\%$)

Number of observations between $8.36 - 3(0.46) = 6.98$ and $8.36 + 3(0.46) = 9.74$ is 39 ($39/39 = 100\%$)

The agreement with the 68–95–99.7 rule is quite good.

Exercise 1.67

Here are the five-number summaries:

DiMaggio
Minimum = 12
1st quartile = 21
Median = 30
3rd quartile = 32
Maximum = 46

Mantle
Minimum = 12
1st quartile = 21
Median = 28.5
3rd quartile = 37
Maximum = 54

Here are side-by-side boxplots:

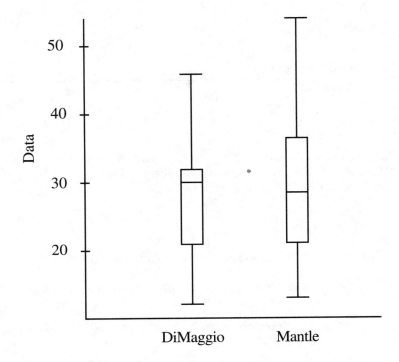

The main difference in the distributions is at the high end, in the upper quartile. Mantle had more years with 40 or more homeruns than did DiMaggio. The distributions of the lower three quarters of the data for both players are similar.

Exercise 1.69

(a) A histogram of the data is:

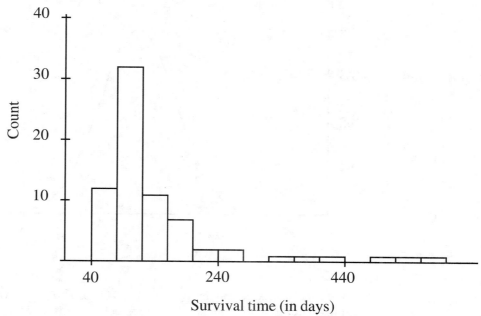

The distribution is strongly skewed to the right with a long tail to the right. There is a marked peak at the bar covering 80 to 120. The six observations above 128 are quite noticeable.

(b) The mean is considerably larger than the median. This is consistent with the strong right skewness of the distribution of the data. The long right tail pulls up the mean.

(c) The five-number summary is (read off the Data Desk printout in the problem):

Minimum = 43
1st quartile = 82.25
Median = 102.5
3rd quartile = 153.75
Maximum = 598

Exercise 1.77

State the problem. We want the proportion of the population used to develop the test with scores below 1.7, i.e., $x \leq 1.7$.

$$\textit{Standardize:} \qquad x \quad \leq \quad 1.\overline{7}$$

$$\frac{x - 3.0}{0.8} \quad \leq \quad \frac{1.7 - 3.0}{0.8}$$

$$z \quad \leq \quad -1.63$$

This corresponds to the following area under the normal curve.

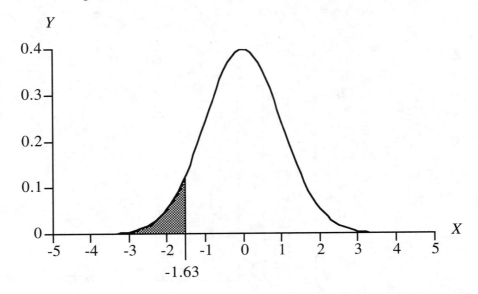

-1.63

Use the table. Area to the left of $-1.63 = 0.0516$. Thus the proportion of the population used to develop the test with scores below 1.7 is 0.0516.

State the problem. We want the proportion of the population used to develop the test with scores between 1.7 and 2.1, i.e., $1.7 \le x \le 2.1$.

$$\textit{Standardize:} \qquad 1.7 \quad \le \quad x \quad \le \quad 2.1$$

$$\frac{1.7 - 3.0}{0.8} \le \frac{x - 3.0}{0.8} \le \frac{2.1 - 3.0}{0.8}$$

$$-1.63 \quad \le \quad z \quad \le \quad -1.13$$

This corresponds to the following area under the normal curve.

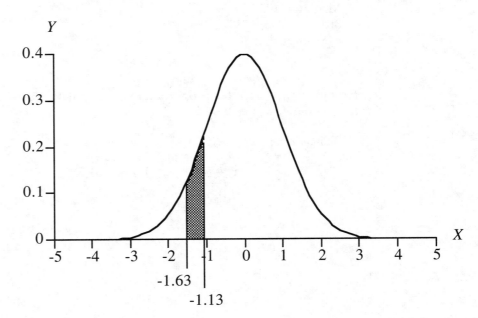

-1.63

-1.13

Use the table. Area between -1.63 and -1.17 = area to the left of -1.17 –area to the left of -1.63
$$= 0.1292 - 0.0516$$
$$= 0.0776$$

Thus the proportion of the population used to develop the test with scores between 1.7 and 2.1 is 0.0776.

CASE STUDY

In assessing the quality of any conclusion drawn from data, it is important to know several things. These include background information such as what variables were measured, how these variables are defined, and how the data were collected. In addition, *one should actually look at the data oneself*, if possible. Since many important policy decisions and laws affecting our lives are based on conclusions drawn from data, it is in everyone's interest to be cautious of unsubstantiated conclusions drawn from data. It is both surprising and frightening how often people cite conclusions based on data without any attempt at substantiating these conclusions. Examples abound in the medical literature. This case study investigates another instance that has made its way into the national media and has been cited as evidence against gun control.

Many people have hypothesized that civilian gun ownership or restrictions on gun ownership may affect the crime rate in the United States. Some say that regulating gun ownership will result in fewer crimes, since there would be fewer guns with which to commit the crimes. Others say that restricting gun ownership would only make it easier for the criminal to commit crimes. They say gun ownership by "law abiding citizens" will discourage the criminal and the crime rate will decrease. Several cities in the U.S. have passed ordinances or laws regulating the ownership of guns. Kennesaw, Georgia passed such a city ordinance on March 15, 1982. The ordinance required every household to own a firearm and became effective in the month of April 1982. According to a city council member this law was intended to express disdain for a law passed in Morton Grove, Illinois, which prohibited the possession or sale of handguns within the city.

Was the law effective in reducing crime? A study in 1989 (Loftin, C., Mcdowall, D., and Wiersema, B., "Did Mandatory Firearm Ownership in Kennesaw Really Prevent Burglaries?", Sociology and Social Research, 74, 1, 48–51.), reports that one investigator compared the number of burglaries in the first seven months following passage of the law (April–Oct. 1982) with the number of burglaries in the same period in the previous year (April–Oct. 1981). This investigator found an 89% decrease in burglaries. The data used by this investigator, which was the basis for the figure of 89%, was from a memorandum of a telephone conversation with the Kennesaw, Georgia Police Chief.

According to a published report in *The Columbus Dispatch* of Feb. 14, 1993, a bill requiring each home in Oregon to have at least one firearm and ammunition on hand was introduced in the Oregon Legislature by state representative Liz Van Leeuwen. The article goes on to say:

> "Van Leeuwen, 67, introduced the bill Jan. 29 at the urging of a gun
> dealer. Patterned after a city ordinance passed 12 years ago in
> Kennesaw, Georgia, it requires gun safety training and exempts
> those who oppose gun possession. The city gun ordinance was
> passed in Kennesaw, Georgia in 1981. Burglaries initially dropped
> by 65%, to 19% in 1982, but then began climbing."

In Dublin, Ohio in 1993, a bill requiring a waiting period for purchase of a gun came before the city council. During public discussion of the proposed bill, the success of the ordinance in Kennesaw, Georgia was cited as a reason against the bill and in favor of gun ownership. In a major TV network news show on guns aired in the Spring of 1994, the success of the Kennesaw, Georgia bill in reducing crime was again referred to as a reason for ownership of guns.

The examples cited in the previous three paragraphs are presumably based on data. They suggest data collected in Kennesaw Georgia show that burglaries dropped substantially after enactment of the ordinance. We can check these claims by examining the data ourselves. In fact, the number of burglaries committed in Kennesaw, Georgia is available from the Uniform Crime Reporting Section of the FBI. These data are given below.

Months are measured from March 1982 (the month prior to the ordinance). Negative signs indicate months before March 1982 and plus signs indicate months after March 1982.

Month	Burglaries	Month	Burglaries
−75	9	0	0
−74	0	1	6
−73	3	2	1
−72	0	3	4
−71	5	4	4
−70	4	5	6
−69	3	6	2
−68	2	7	3
−67	4	8	0
−66	7	9	2
−65	2	10	5
−64	2	11	1
−63	1	12	0
−62	2	13	3
−61	0	14	2
−60	3	15	5
−59	4	16	5
−58	1	17	2
−57	0	18	4
−56	0	19	2
−55	0	20	4
−54	0	21	1
−53	4	22	2
−52	6	23	2
−51	1	24	0
−50	0	25	3
−49	1	26	3
−48	4	27	4
−47	0	28	4
−46	4	29	4
−45	2	30	3
−44	2	31	2
−43	2	32	1
−42	2	33	3
−41	1	34	3
−40	3	35	2
−39	3	36	1
−38	1	37	6

Month	Burglaries		Month	Burglaries	(Continued)
−33	2		42	2	
−32	7		43	4	
−31	0		44	1	
−30	0		45	2	
−29	0		46	4	
−28	0		47	2	
−27	3		48	8	
−26	1		49	10	
−25	6		50	6	
−24	6		51	9	
−23	3		52	7	
−22	7		53	2	
−21	2		54	5	
−20	7		55	5	
−19	0		56	10	
−18	0				
−17	0				
−16	0				
−15	3				
−14	4				
−13	9				
−12	4				
−11	7				
−10	10				
−9	6				
−8	5				
−7	1				
−6	4				
−5	1				
−4	0				
−3	1				
−2	6				
−1	2				

We begin by calculating the total number of burglaries in the first seven months following passage of the law (April–Oct. 1982) and in the same period in the previous year (April–Oct. 1981). For April–Oct. 1982 (months 1 to 7) the total number of burglaries is calculated to be 26. The total for the same period in the previous year (months −11 to −5) is 34. The reduction is (34 −26)/34 × 100% = 23.5%. This is more than a third less than the reported reduction of 89%! Although we are unable to produce the figure of 89%, this is still a reduction. Isn't this evidence of a drop in crime? One response is to ask why the April–October periods were selected. If we examine the 6 months before (−6 to −1) and after (1 to 6) passage of the ordinance, we find 14 burglaries in the 6 months prior to passage and 23 in the 6 months after passage. This appears to show that that the ordinance increased burglaries! Examining selected brief periods before and after passage of the ordinance does not appear to be a convincing way to assess the impact of the law.

It is also not clear where the statement "Burglaries initially dropped by 65%, to 19% in 1982, but then began climbing," which appears in *The Columbus Dispatch* article, comes from. In fact, what does the 19% refer to? 19% of what? These numbers are without context and impossible to reproduce or interpret without more information.

Perhaps a more fruitful way to examine the data is graphically with a time plot. A time plot of the data is therefore given below. Dots indicate months prior to passage of the ordinance, x's months after passage of the ordinance.

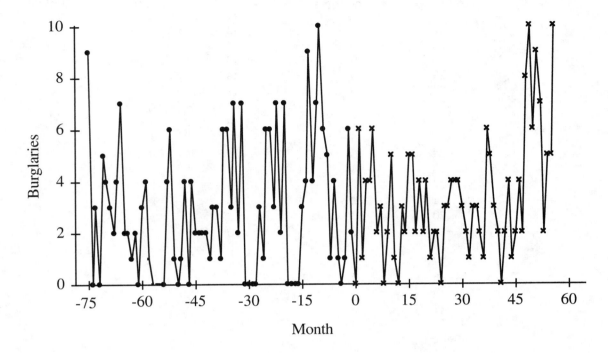

Month

What does the time plot show? The peak from −13 to −10 months gives an initial impression of a drop in burglaries at about the time the ordinance was passed. However the actual drop in burglaries occurred prior to passage of the ordinance. Similar peaks, follwed by dips in burglaries occur at −75 and at 55. These cannot be related to the ordinance. We note in passing that these peaks appear to occur at about five-year (60 month) intervals. If we had more data, it would be interesting to see if this five-year pattern persists. Even more interesting would be to attempt to account for the pattern.

If one looks hard at the plot, there does appear to be a general decline in burglaries for a period of about 30 months after passage of the ordinance as compared to the 30 months prior to passage but this seems to be part of the long dips between the three peaks. One could just as easily note the rise in crime starting at about 40 months after passage of the ordinance. Should this be interpreted as a crime spree due to the proliferation of guns in the community? Notice also that there were more runs of consecutive months without burglaries prior to passage of the ordinance than after passage. At best, then, there is very weak evidence of a reduction in burglaries after passage of the ordinance when the actual data are examined over a long period. Certainly, it is not clear where the figure of an 89% drop in burglaries (reported by the Chief of Police) came from. We were not able to produce this number from the actual data by any reasonable method.

Here is an instance where the numbers cited in the media do not seem to be supported by the data. Looking at all the data (not just selected summary statistics or small portions of the data) helps us assess the actual impact of the ordinance. The data suggest that, at best, the law had little effect, and certainly no lasting effect, on reducing the number of burglaries. They neither decreased nor increased consistently after passage of the ordinance. It does not seem reasonable to use Kennesaw, Georgia as the basis for any laws or policy decisions regarding gun control, pro or con. Unfortunately, the media continues to use Kennesaw, Georgia as an example of the effectiveness of gun ownership on controlling crime!

We hope the lesson is clear, you need to actually look at the data in order to assess arguments based on data.

CHAPTER 2

EXAMINING RELATIONSHIPS

CHAPTER OVERVIEW

Now we have the tools to explore several types of variables one by one, but it is very rare that a survey only asks one question or a researcher only looks for one kind of result. Data are most often a collection of variables that may have some kind of relationship. This relationship may be more interesting than any of the variables alone. Does the number of hours a student sleeps have any relationship the number of hours a student studies? Is there a connection between the weight of a car and its MG (miles per gallon)? In Chapter 2 we will examine ways of exploring these relationships. The chapter concentrates on the relationships of quantitative variables, but categorical variables are discussed in the last section.

As in Chapter 1, the analysis of multiple variable data has several steps:

- Begin by looking at the data. First, we need to understand what the variables are. Ask yourself some basic questions about the data and what you want to learn from them. Are the variables quantitative or categorical? Is one variable possibly causing another?

- Next, try some graphs. Chapter 2 introduces new graphs such as the scatterplot and shows new ways of using the ones we met in the first chapter.

- After the graphs, we can look at numerical summaries of the data. The mean, median, standard deviation, and quartiles still have the same meanings. Now we also have ways of exploring the relationships between the variables: correlation and least-squares regression, to name two.

- Finally, there are some cautions to keep in mind because these relationships are not always clear cut.

SECTION 2.1

SECTION OVERVIEW

The first tools we have for examining variables are graphical. Histograms, pie charts and box plots work very well for single variables. For more than one variable we have the scatterplot. Scatterplots show us two quantitative variables at a time. A third, categorical, variable can be added by using colors or different symbols. As with histograms, we will look for **direction, form, strength,** and of course outliers in scatterplots. The patterns may tell us about any connection between the variables. Is there any sort of **association** between the **response variable** and the **explanatory variable**?

KEY CONCEPTS

How to draw a scatterplot

1. List the variables so you can see how the values of the variables are matched to each other.

   ➤ Here are acorn sizes, in cm³, and eventual tree heights, in meters.

Species	Acorn size	Tree height
a	1.4	27
b	3.4	21
c	9.1	25
d	1.6	3
e	10.5	24
f	2.5	17
g	0.9	15
h	6.8	.30
i	1.8	24
j	0.3	11

2. Pick one variable to be on the horizontal axis and put the other on the vertical axis. Usually the explanatory variable is placed on the horizontal axis.

3. Plot each individual variable on the graph. Make sure to match the values of the variables to the correct individuals.

➢ Here is a scatterplot of tree heights and acorn sizes.

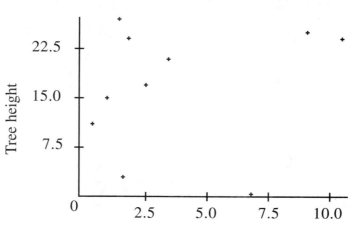

Acorn size

How to use a scatterplot

Use a scatterplot to look for patterns. Do the data show any association? **Positive association** is when the variables both take on high values together. **Negative association** is when one variable takes high values and the other takes on low values.

A strong association often makes the variables appear to have a **linear relationship**. The plotted values seem to form a line. If the line slopes up to the right the association is positive. If the line seems to slope down to the right, the association is negative.

As always, look for outliers. The outlier may be far away in terms of the horizontal variable or the vertical variable.

SOLUTIONS TO SELECTED TEXT EXERCES

Exercise 2.3

The response variable is how long a patient lived after treatment, since it is the outcome of the study. Since it is a measure of time, it is a quantitative variable.

The explanatory variable is which treatment (the removal of the breast vs. the removal of only the tumor and nearby lymph nodes) a patient received. The treatments attempt to explain the response (how long after treatment a patient lived). This is categorical since it is not a number.

Exercise 2.7

(a) Since the researchers believe that lean body mass is an important influence on (and hence helps explain) metabolic rate, we treat lean body mass as the explanatory variable and metabolic rate as the response. Here is a scatterplot of metabolic rate vs. lean body mass for the female subjects.

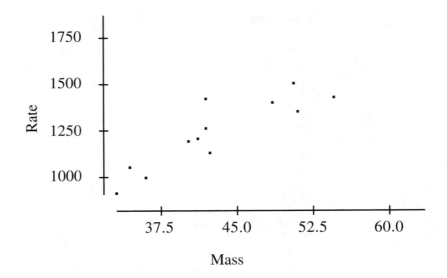

(b) As lean body mass increases, or as you move from left to right across the horizontal axis in the scatterplot, the points in the plot tend to rise. This indicates that the association between the variables is positive. The form of the relationship appears to be linear since a straight line seems to be a reasonable approximation to the overall trend in the plot. The relationship is not perfect but appears to be moderately strong.

(c) We add the men to the plot. Men are indicated by the x's.

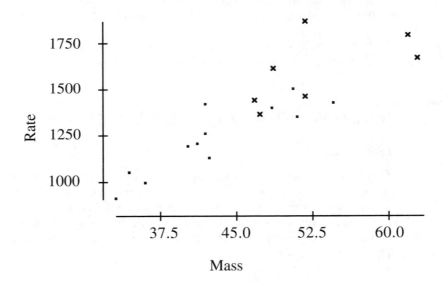

The male subjects also show a positive association that might be described as linear. The association does not appear as strong as for the women and the slope of the linear relation may be a bit flatter. We also notice that the men are clustered in the upper right of the plot. This is not surprising, since men tend to be larger than women.

Exercise 2.11

For these data we do not envision one of the variables as explaining the other. All we are interested in is investigating the association between the two variables. We are free to arbitrarily designate one of the variables as the explanatory variable and the other as the response. We will choose the variable femur as the explanatory variable, plotting it on the horizontal axis, and choose humerus to play the role of the response, plotting it on the vertical axis. The resulting scatterplot is given below.

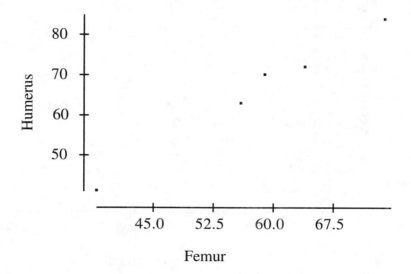

As expected the association is positive. It is also strongly linear. One of the points (the third point with Femur = 59 and Humerus = 70) appears to differ a bit from the others. If this point is ignored, the other four appear to lie nearly exactly on a straight line. While the difference does not appear to be dramatic, this point might come from a different species than the others. The evidence does not appear overwhelming, however.

Exercise 2.13

(a) Part of what states spend on education goes to teacher salaries. Thus those states that spend more on education should tend to have higher teacher salaries. A related factor that would lead us to expect a positive association between what states spend on education and teacher salaries is the cost of living. Areas, such as urban areas, with a higher cost of living will have to spend more on education and pay teachers more.

(b) A scatterplot with education spending (dollars per pupil) as the explanatory variable is given below.

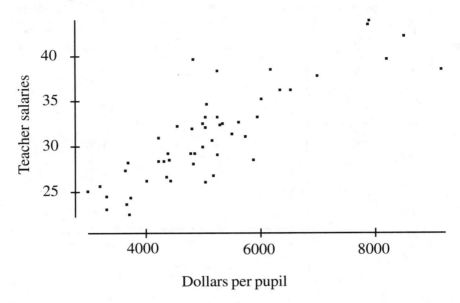

Dollars per pupil

(c) The scatterplot does show a positive association, with teacher salaries tending to be higher in states with higher spending per pupil. The association also is approximately linear.

(d) The state indicated by the + symbol in the plot below is one with unusually high teacher salaries relative to the state's education spending. If we check which state it is, we find it is California.

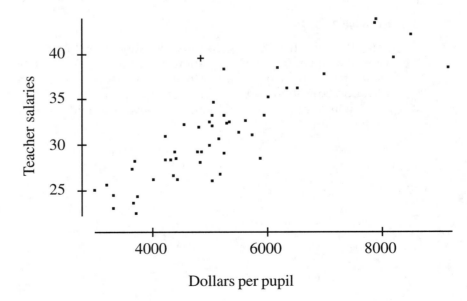

Dollars per pupil

(e) The mountain states are indicated in the scatterplot below by x's.

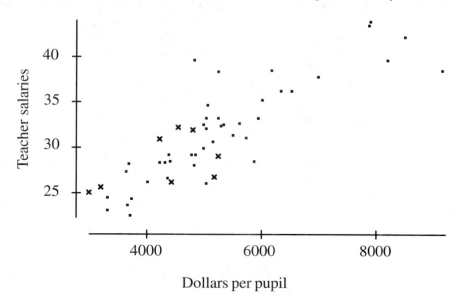

Notice the mountain states are in the lower left portion of the plot. Generally the mountain states tend to spend below average amounts on education and to pay below average teacher salaries than other states.

SECTION 2.2

SECTION OVERVIEW

Now that we know how to look at variables, we may want to use a numerical summary to describe the relationships we noticed in the scatterplots. Our eyes are good tools for many things but a summary number is often a better way to describe the associations our eyes may not catch. **Correlation** is the measure we will use to show the strength of **linear association**.

KEY CONCEPTS

How to calculate correlation

You should have a calculator (*and are strongly encouraged to purchase one!*) which will calculate the correlation r and the least squares line directly. The correlation can be calculated relatively easily using even a simple calculator as follows.

1. From your calculator obtain the mean for each variable, that is $\bar{x}$ and $\bar{y}$.

2. From your calculator obtain the standard deviation for each variable, that is s_x and s_y.

3. From these pieces calculate r, using $r = \dfrac{1}{n-1} \sum (\dfrac{x_i - \bar{x}}{s_x})(\dfrac{y_i - \bar{y}}{s_y})$

REMEMBER:

- If **r** *is positive it means that there is a positive association.*

- If **r** *is negative it means that there is a negative association.*

- **r** *will not change if we switch x and y.*

- **r** *is always between 1 and –1. Values close to 1 or –1 show strong association. Values near 0 show weak association.*

- **r** *is strongly affected by outliers. Outliers can make the correlation much different than what the correlation might be if the outliers are removed.*

- *Correlation measures straight line association. Any other patterns will not be captured by the correlation coefficient.*

SOLUTIONS TO SELECTED TEXT EXERCISES

Exercise 2.17

(a) Let x denote the femur measurements and y the humerus measurements. We find the means and standard deviations to be

$$\bar{x} = 58.20 \qquad s_x = 13.1985$$

$$\bar{y} = 66.00 \qquad s_y = 15.8902$$

We summarize the calculations for the correlation r in the following table

x	$\dfrac{x - \bar{x}}{s_x}$	y	$\left(\dfrac{y - \bar{y}}{s_y}\right)$	$\left(\dfrac{x - \bar{x}}{s_x}\right)\left(\dfrac{y - \bar{y}}{s_y}\right)$
38	−1.5305	41	−1.5733	2.4079
56	−0.1667	63	−0.1888	0.0315
59	0.0606	70	0.2517	0.0153
64	0.4394	72	0.3776	0.1659
74	1.1971	84	1.1328	1.3561

The sum of the values in the last column above is 3.9767. Thus the correlation is

$$r = 3.9767/4 = 0.9942.$$

(b) Our calculator (actually our software) gives a correlation of 0.994. This agrees with the result in (a).

Exercise 2.21

(a) Refer to the solutions to Exercises 2.11 and 2.17 above. The points on the scatterplot lie quite close to a positively sloping straight line, which leads us to expect a correlation near 1. The correlation calculated in Exercise 2.17 was 0.9942, quite close to 1 as suggested by the scatterplot.

(b) Changing the units of measurement from centimeters to inches will not change the correlation. Remember, the formula for correlation involves the standardized values of the variables that remain unchanged by a change in the units of measurement.

Exercise 2.25

(a) Here is a scatterplot of the data.

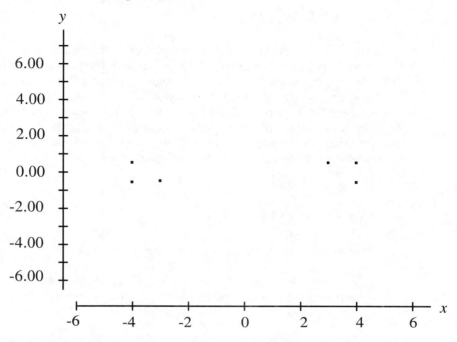

(b) Here is the plot in (a) with the x^* and y^* points added as x's.

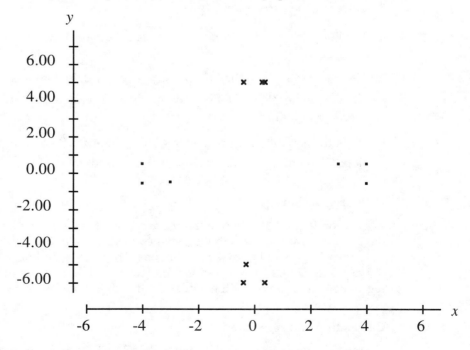

(c) The correlation between x and y is 0.253 (as calculated by our software). The correlation between x^* and y^* is also 0.253. The two correlations are the same. This is not surprising since x^* is a multiple (1/10) of x and y^* is a multiple (10) of y. This means that both x^* and x will have the same standardized values. Likewise, y^* and y will also have the same standardized values. As a result, the correlation between x and y, and between x^* and y^* will be the same.

SECTION 2.3

SECTION OVERVIEW

After plotting the data in a scatterplot and deciding that the correlation is strong enough to merit deeper exploration, we may further describe the data using **least-squares regression.** Least-squares regression fits a line to the data. When the conditions are right, this regression method is a good summary of the relationship between the variables. The regression line can even be used to **predict** the value of additional values of the response variable. A form of the correlation coefficient, r^2, is a measure of the goodness of the regression. An examination of the **residuals** shows us how well our regression does in predictions.

KEY CONCEPTS

How to calculate the least-squares regression line

1. You must decide which variable is the explanatory variable and which is the response variable. In regression the line calculated will be different if you switch the variables.

2. Calculate the means, $\bar{x}$ and $\bar{y}$, and the standard deviations, s_x and s_y.

3. Calculate the correlation, r.

4. The line will have the slope $b = r\frac{s_y}{s_x}$. Slope is the amount y changes when x increases by 1.

5. The intercept, where the line crosses the vertical, or y, axis is a $= \bar{y} - b\bar{x}$.
   The line is completely defined by the slope and intercept. With these two pieces we can plot a line.

6. Write the equation for the line: $\hat{y} + b\,x$, where a is the intercept and b is the slope. Remember that $\hat{y}$ means "the predicted value of y."

Using regression

Regression is used to describe the connection between the explanatory variable and the response variable. It is the line that comes closest to hitting all the points.

Regression is used to make predictions. The regression line is the line that marks the average value of y, given a value of x. If we take a value of x and put it in the regression equation we will get back a value of y, called $\hat{y}$.

Correlation and regression

Correlation and regression are obviously related. Just look at the form of the slope, b. However, the more important connection is how r^2, the square of the correlation coefficient, measures the strength of the regression. r^2 tells us how much of the variation in y is explained by the regression of y on x. The closer r^2 is to 1 the better the regression describes the connection between x and y.

Residuals

The difference between an observed value of y and the predicted value obtained by least-squares regression is called the residual.

$$\text{residual} = y - \hat{y}$$

We use residuals to check the usefulness of our regression.

Residual plots

1. Calculate, or otherwise obtain, the residual values.

2. Make a scatterplot of the residuals against the explanatory variable.

3. Draw a horizontal line at zero. This makes any patterns easier to see.

4. Look for patterns and outliers. Any sort of pattern may indicate that regression is not a good tool to use, that the relationship may not be linear. Outliers may signify an influential observation that is affecting the whole regression line.

➤ The following scatterplot is an example of a problem. The points are clustered near the low end with a number of outliers high along the horizontal axis. These points may be influencing the regression too much. Try a regression with them removed.

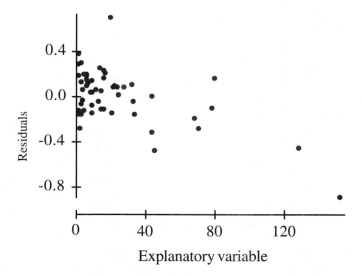

SOLUTIONS TO SELECTED TEXT EXERCISES

Exercise 2.29

(a) Our software gives the following equation of the least-squares regression line.

$$\hat{y} = 1.08921 + 0.188999x$$

(b) Our software gives the following statistics.

$$\bar{x} = 22.3125 \qquad s_x = 17.7378$$

$$\bar{y} = 5.3063 \qquad s_y = 3.3684$$

$$r = 0.995$$

We calculate

$$b = r(s_y/s_x) = 0.995(3.3684/17.7378) = 0.18895$$

$$a = \bar{y} - b\bar{x} = 5.3063 - 0.18895(22.3125) = 1.09035$$

Both these results and those of (a) agree with each other, taking into account round-off error which will make results differ in the third or fourth decimal place.

Exercise 2.33

(a) Here is a scatterplot with speed as the explanatory variable and steps per second as the response variable.

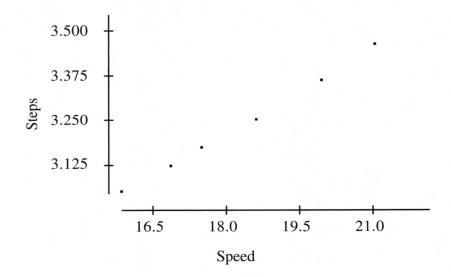

(b) The data show a strong linear, positive association. The points appear to come close to lying on a positively sloping straight line. We calculate the correlation to be

$$r = 0.999$$

(c) Our software gives the equation of the least-squares regression line as

$$\hat{y} = 1.766 + 0.080x$$

where y represents steps per second and x running speed. If we add this regression line to our scatterplot we get the plot

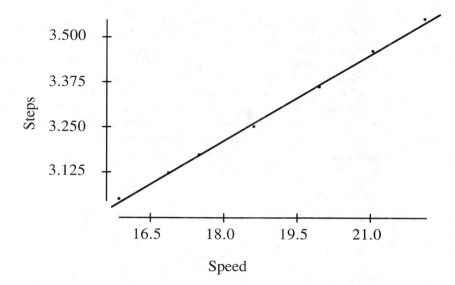

(d) We calculate $r^2 = (0.999)^2 = 0.998$. Thus running speed explains 99.8% of the variation in number of steps that a runner takes per second. This is obviously most of the variation!

(e) These data are for top female runners. The regression line is really only valid for such runners. If we use it to predict running speed from average steps per second for other runners, our prediction is likely to be less accurate. Extending the regression to include a wider variety of runners will introduce other factors affecting the variation of running speed into the data, such as the sex of the runner, greater variation in body size, and skill level. This should cause r^2 to decrease.

Exercise 2.37

(a) If we leave out Child 19, we obtain (from our software or calculator) the following least-squares regression line.

$$\hat{y} = 109.305 - 1.193x$$

Here is a graph of the regression lines with and without Child 19.

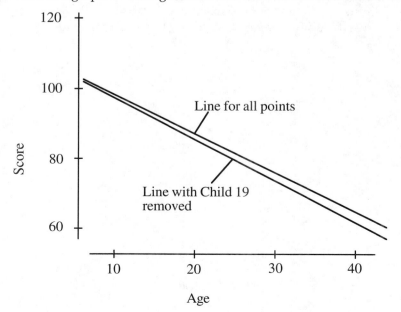

Although removing Child 19 does cause some change in the least-squares regression line (the line shift downward slightly), the change is not dramatic. We would not call Child 19 very influential.

(b) With Child 19 in the data, $r^2 = 0.410$. With Child 19 removed, $r^2 = 0.572$. r^2 increases when we remove Child 19 from the data. As an outlier, Child 19 adds to the variability of the data about the regression line. The data appear more tightly clustered about the least-squares regression line with Child 19 removed than when Child 19 is kept in the data. The more tightly clustered the data is about the least-squares regression line, the closer the correlation is to $+1$ or -1 and the larger the value of r^2. Thus removal of Child 19 should increase r^2.

Exercise 2.45

(a) Below is a scatterplot of the data with the regression line drawn on it.

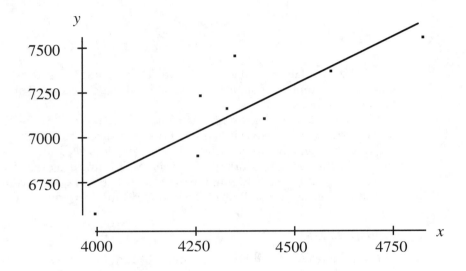

The percent of the variation in class enrollment that is explained by the linear relationship with the count of freshmen is determined by r^2. Here, $r^2 = (0.8333)^2 = 0.6944$. Thus 69.44% of the variation in class enrollment that is explained by the linear relationship with the count of freshmen.

(b) If we sum the residuals, we get (using a calculator) exactly 0!

(c) A plot of the residuals against year is given below.

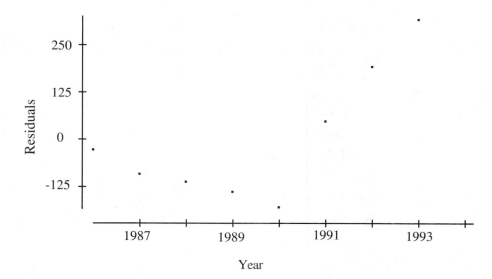

There is a change in the trend of the residuals beginning in 1991. Prior to 1991 the residuals steadily get more negative, indicating that the least-squares regression line was steadily overestimating the number of students signing up for freshman-level math courses. Beginning in 1991 the residuals become increasingly positive, indicating that the least-squares regression line was steadily underestimating the number of students signing up for freshman-level math courses. This is consistent with the change in program described, which would create an increase in the enrollment in math courses. This suggests that the change occurred in 1991.

SECTION 2.4

SECTION OVERVIEW

Section 2.4 is the warning label. In the first parts of the chapter we learned some new and potentially powerful ways of describing the connection between two variables. Now we need to watch the limitations. In summary, here is what to watch out for:

- Do not **extrapolate** beyond the range of the data.

- Be aware of possible **lurking variables**.

- Are the data averages or from individuals? **Averaged data** usually lead to overestimating the correlations.

- Most of all, remember that *association is not causation*!! Just because two variables are correlated doesn't mean one causes the other.

SOLUTIONS TO SELECTED TEXT EXERCISES

Exercise 2.47

(a) Here is a scatterplot of the data with year the explanatory variable and population the response.

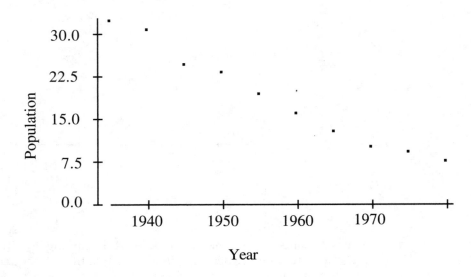

We find that the least-squares regression line is

$$\hat{y} = 1166.93 - 0.58679x$$

(b) The slope of the least-squares regression line indicates the decline in farm population per year over the period represented by the data. This is a decline of 0.5868 million people per year or 586,800 people per year. The percent of the observed variation in farm population accounted for by linear change over time is determined by the value of r^2, which is 0.977 (calculated using software). The desired percent is therefore 97.7%.

(c) In 1990 the regression equation predicts the number of people living on farms to be

$$\hat{y} = 1166.93 - 0.58679(1990) = -0.782 \text{ million.}$$

This result is unreasonable since population cannot be negative.

Exercise 2.49

We would expect the correlation to be lower. In general, *correlations based on averages are usually too high when applied to individuals.* Averages remove individual to individual variation. This additional variation generally tends to reduce the correlation.

Exercise 2.51

Both shoe size and score on a reading comprehension test should increase as a child ages. The positive correlation is not surprising and is explained by the lurking variable age.

Exercise 2.55

The explanatory variable is foreign language study. It has two values depending on whether a student has studied at least two years of a foreign language or has studied no foreign language. The response is the score on the English achievement test administered to seniors. Unfortunately for the study, students choose whether or not to take a foreign language. Academically stronger students are more likely to study a foreign language than weaker students. The academic strength of a student is the lurking variable here and prevents us from concluding that language study improves student's English scores.

SECTION 2.5

SECTION OVERVIEW

In the previous sections the tools were for use with quantitative variables. Now we will look at some ways of handling non-numerical data. The primary tool is the **two-way table**, which is an easy way to list the counts, or percents, of individuals that fall into each category. As for graphical tools we can use a **bar graph**, which looks suspiciously like a histogram. Finally, there are some warnings about the information in the two-way tables. Like lurking variables, **Simpson's paradox** can muddle up the results quite thoroughly.

KEY CONCEPTS

Making a two-way table

1. Write all the possible classes of each of the two categorical variables. Write one list horizontally and the other vertically.

   ➤ Your data are the age group a man falls in and whether or not he has a mustache

	Mustache	
Age group	Yes	No
17–24		
25–35		
36–45		
46–60		
60+		

2. To complete the table, write in the counts of the individuals that fall into each category. Sum up each row and column.

➤ Continued from the previous page.

| | Mustache | | |
Age group	Yes	No	Total
17–24	12	65	77
25–35	22	45	67
36–45	42	30	72
46–60	30	38	68
60+	32	46	78
Total	138	224	362

3. To use the table look at the marginal distributions of the row and column variables. Calculate the percents that will help you answer whatever question you want to answer. Percents are more descriptive than just the counts.

   ➤ What percent of men between the ages of 36 and 45 have mustaches?

   There are 72 men in this category. Forty-two have mustaches. This means that about 58% of men ages 36–45 have mustaches. (The calculations: $42 \div 72 = .5833$ and so the percent is $.5833 \times 100 = 58\%$.)

How to make a bar graph

1. Draw the vertical scale to represent the percents. Mark out a bar to represent each category of the variable the chart is for.

2. Draw the bars to the height equal to the percent of the observations that fit in that category.

3. The finished bar graph will look like a histogram but there should be spaces between the bars, and the horizontal axis should have no units of measurement.

SOLUTIONS TO SELECTED TEXT EXERCISES

Exercise 2.59

The marginal distribution of age can be found from the totals in each age group given in the bottom row of Table 2.5. Each value in this row must be divided by the total number of persons represented by the table found in the lower right corner, which is 158,694 thousand. The result is

	Age Group				
	25–34	35–44	45–54	55–64	≥65
Fractions	$\dfrac{42,905}{158,694}$	$\dfrac{38,665}{158,694}$	$\dfrac{25,686}{158,694}$	$\dfrac{21,346}{158,694}$	$\dfrac{30,092}{158,694}$
Percent	27.04%	24.36%	16.19%	13.45%	18.96%

Exercise 2.61

The percent of people in each age group who did not complete high school is determined by dividing the counts in the "Did not complete high school" row of the table by the total for this row (which is 34,251). The result is

Age Group

	25–34	35–44	45–54	55–64	≥65
Fractions	$\dfrac{5965}{42,905}$	$\dfrac{4755}{38,665}$	$\dfrac{4829}{25,686}$	$\dfrac{5999}{21,346}$	$\dfrac{12,702}{30,092}$
Percent	13.90%	12.30%	18.80%	28.10%	42.21%

A bar graph displaying these percents is given below.

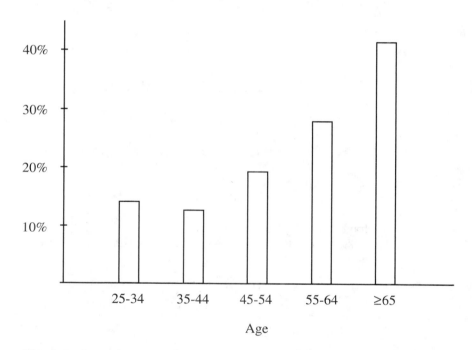

The data show that a much greater percentage of people over 65 did not complete high school than in the other age groups. The other age groups are similar with slightly higher percentages in the 55–64 and 25–34 groups. One possible explanation is that a greater emphasis on education (completion of high school) existed beginning in the 1950s (with concerns about the cold war and the space race) and this is reflected in the increasing percentage of people who did not complete high school with age. The larger percentage in the 25–34 age group may reflect a growing dropout rate in recent years.

Exercise 2.66

(a) The total number of men who took part in the study can be calculated by adding all four entries in the table. The result is 6014. The number who died during the study is $21 + 55 = 76$, the total of the entries in the first column. The percent who died during the five years of the study is therefore

$$\% \text{ died} = 76/6014 \times 100\% = 1.26\%$$

(b) Blood pressure is the explanatory variable. We want to explain outcome (died or survived) using blood pressure.

(c) In order to determine if high blood pressure is associated with a higher death rate, we calculate the death and survival percentages in each row. The total in the low blood pressure row is 2676 and in the high blood pressure row is 3338. Dividing the entries in each row by the corresponding row total and converting to percents gives the following.

	Died	Survived
Low blood pressure	0.78%	99.22%
High blood pressure	1.65%	98.35%

This indicates that the death rate in the high blood pressure group is about twice that in the low blood pressure group and suggests there is an association between high blood pressure and a higher death rate.

Exercise 2.70

(a) Here is a two-way table of player versus outcome.

	Hits	At bats
Joe	120	500
Moe	130	500

(b) The overall batting averages (hits/at bats) for each player is

Joe = 120/500 = .240
Moe = 130/500 = .260

Moe has the higher batting average.

(c) Here are separate two-way tables for each player versus outcome for each kind of pitcher.

	Right		Left	
	Hits	At bats	Hits	At bats
Joe	40	100	80	400
Moe	120	400	10	100

From these two tables we calculate the following batting averages

	Right	Left
Joe	40/100 = .400	80/400 = .200
Moe	120/400 = .300	10/100 = .100

Joe has a batting average .100 higher than Moe against both right- and left-handed pitching.

(d) Both Joe and Moe bat much better against right-handed pitching than left-handed pitching. Thus the more either one bats against left-handed pitching, as compared to right-handed pitching, the worse their overall batting average will be. Joe bats much more often against left-handed pitching than Moe, hence his overall average suffers more than Moe's. Therefore even though Joe bats better than Moe against both right- and left-handed pitching, since he bats much more often against left-handed pitching than Moe, his overall average is lower.

Exercise 2.73

(a) The number of undergraduates enrolled in colleges and universities can be obtained by adding all the entries in the last row (the total row) of the table. This yields a total of

$$2212 + 2065 + 5657 + 1440 = 11{,}374 \text{ thousand}$$

or 11,374,000 students.

(b) The percent of all undergraduates that were 18–21 years old in the fall of the academic year is obtained by adding up the entries in the 18–21 row and dividing by the total number of undergraduates calculated in (a). The total of the 18–21 row is

$$1345 + 456 + 3869 + 159 = 5829$$

so the appropriate percent is 5829/11,374 ¥ 100% = 51.25%.

(c) The percent of undergraduate students enrolled in each of the four types of program who were 18–21 is obtained by dividing the 18–21 year old entry in each column by the total for the column and then converting to a percent. If we do this we obtain

	2-year full-time	2-year part-time	4-year full-time	4-year part-time
Proportion	1345/2212	456/2065	3869/5657	159/1440
Percent	60.80%	22.08%	68.39%	11.04%

A bar chart of these percents is

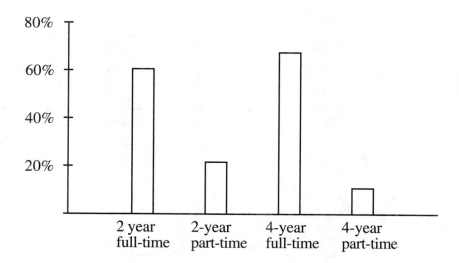

Type of college/university

(d) Only slightly over half of all undergraduate students are in the "traditional" 18–21 age group. The other half are in "non-traditional" groups. The "traditional" age group predominates full time colleges, both two- and four-year institutions. Even in these institutions, this age group constitutes only 60% –70% of the students. They are a minority in part-time institutions. Presumably these part-time institutions appeal to the other age groups that may contain people living at home or people with jobs who can only go to school part-time. Overall, the "traditional" age group, 18–21, is not as predominant as one might expect.

Exercise 2.80

(a) The sum of the entries in the 18–24 row is actually $9008 + 3352 + 8 + 257 = 12,625$. This differs from the entry in the Total column. This discrepancy is undoubtedly due to rounding off of individual table entries to the nearest thousand women.

(b) The distribution of marital status for all adult women is obtained by calculating the individual column totals. Here is what we get.

	Single	Married	Widowed	Divorced
Total	18,541	56,838	11,290	9161

Converting these to percents by dividing by the grand total of 95,833 yields

	Single	Married	Widowed	Divorced
Percent	19.35%	59.31%	11.78%	9.56%

A bar graph of these percents is

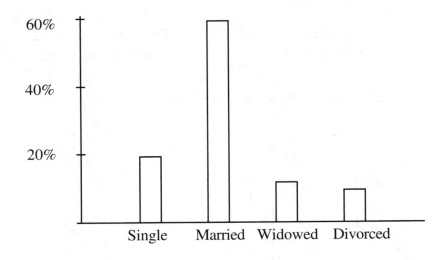

Marital status

(c) We give the distributions, in percents, of marital status for both age groups.

	Single	Married	Widowed	Divorced
18–24	71.34%	26.55%	0.06%	2.04%
40–64	5.85%	72.45%	7.61%	14.08%

The major difference is one of experience with marriage. The majority of women 18–24 (71.34%) are single and have never been married and very few are widowed or divorced. The great majority of women 40–64 (94.14%) are or have been married. These differences probably reflect the fact that marriage, widowhood, and divorce are more likely as one gets older.

(d) The distribution of ages among single women (in percents) is obtained by dividing the entries in the single column by the total for that column (found to be 18,541 in (b)). The result is

	18–24	25–39	40–64	≥65
Percents	48.58%	35.91%	10.65%	4.85%

Obviously, the percent of single women decreases as age increases.

SOLUTIONS TO SELECTED TEXT REVIEW EXERCISES

Exercise 2.86

(a) A scatterplot of assets (the explanatory variable) versus income (the response) is given below with Franklin Bank (bank number 19) indicated by the x.

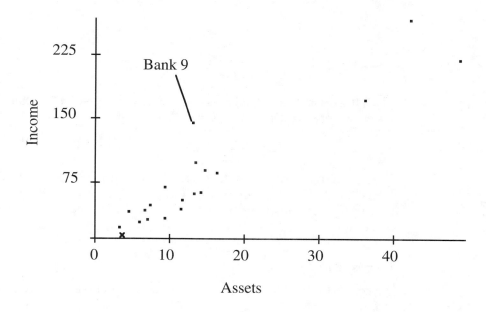

(b) The overall pattern shows a positive association between assets and income. The association appears to be approximately linear. Bank number 9 has an unusually high income (about 144 billion dollars) relative to assets (about 13 billion dollars)and lies above the general trend of the rest of the banks. Three banks (in the upper right of the plot) have both high income and assets, but are reasonably consistent with the trend of the other banks. Franklin Bank only stands out in that it is in the lowest left corner of the plot. Otherwise it is consistent with the trend of the other banks in the plot.

(c) Using our software we find the least-squares regression line for predicting a bank's income (y) from its assets (x) to be

$$\hat{y} = 7.573 + 4.987x$$

The least-squares regression line is plotted below on our scatterplot.

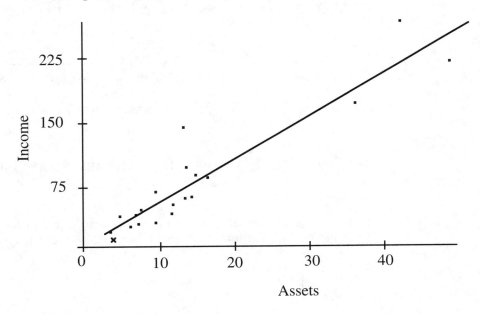

(d) If we use the least-squares regression line to predict Franklin's income from it's assets, we predict it's income to be

$$\hat{y} = 7.573 + 4.987(3.8) = 26.524.$$

The actual income (13.8) was lower than this and the residual is $13.8 - 26.524 = -12.724$. Although the predicted value is almost double the actual value, Franklin Bank lies reasonably close to the least-squares regression line (see the plot in (c)).

Exercise 2.88

We convert the table to percents of males and females by dividing by the column totals.

The new table is

	Male	Female
Firearms	65.87%	42.06%
Poison	13.03%	35.64%
Hanging	14.92%	12.23%
Other	6.19%	10.08%

A few things are striking. About four times as many men (24,724) as women (6182) commit suicide. Our prior belief was that women were more likely to commit suicide than men. Firearms are the most popular method for both sexes, but men use firearms more frequently than women. Poison is nearly as popular as firearms for women. Poison is not as popular for men as it is for women. This may suggest that women have a greater preference for slower, more passive methods (such as poison) than men. Men prefer quicker methods (firearms).

Exercise 2.90

(a) Using computer software we find the correlation between the states' SAT verbal and math scores to be 0.962. If we had the individual scores of many students we would expect the correlation to be lower. Remember, *correlations based on averages are usually too high when applied to individuals.* Averages remove individual to individual variation. This additional variation generally tends to reduce the correlation.

(b) Using a verbal SAT score as the explanatory variable (x) and a math SAT score as the response (y) we obtain the following equation for the least-squares regression line.

$$\hat{y} = 13.823 + 1.079x$$

If a state's median verbal SAT score is 455, the least-squares regression line predicts it's math SAT score to be

$$\hat{y} = 13.823 + 1.079(455) = 504.768$$

(c) A plot of the residuals against the predicted verbal SAT score is given below.

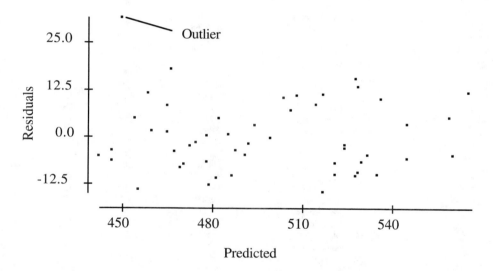

The outlier turns out to be Hawaii. The predicted median math SAT score is about 450, while the actual is 481, higher than predicted on the basis of its median verbal SAT score.

CASE STUDY

Many government agencies and private organizations collect data. These data are studied by researchers in a variety of disciplines. Surprising relationships are sometimes found. The data given below are a very small sample of the data available on the 50 states.

Explanations of the variable names follow the list of data.

State	Pop.(1990)	Low birth weights (under 2500g)	Deaths in fatal accidents	Belt laws 1990	Deaths per 100000 people in fatal accidents
Alabama	4040587	5334	1158	0	28.2
Alaska	550403	575	104	0	19.8
Arizona	3665228	4375	1003	0	28.8
Arkansas	2350725	2986	667	0	27.8
California	29760021	35558	5632	1	19.9
Colorado	3294394	4283	544	1	16.5
Connecticut	3287116	3309	493	1	15.3
Delaware	666168	849	169	0	25.6
Florida	12937926	14813	3,247	1	26.3
Georgia	6478216	9747	1,756	1	27.7
Hawaii	1108229	1445	134	1	14.0
Idaho	1006749	928	278	1	27.7
Illinois	1143602	14785	1,925	1	16.6
Indiana	5544159	5648	1,142	1	20.5
Iowa	2776755	2124	556	1	19.6
Kansas	2477574	2435	494	1	19.8
Kentucky	3685296	3838	843	0	22.6
Louisiana	4219973	6650	946	1	21.5
Maine	1227928	892	268	0	22.2
Maryland	4781468	6229	793	1	17.1
Massachusetts	6016425	5417	750	0	12.7
Michigan	9295297	11586	1,749	1	18.9
Minnesota	4385099	3437	675	1	15.7
Mississippi	2573216	4159	791	0	30.2
Missouri	5117073	5637	1,163	1	22.6
Montana	799065	715	215	1	26.7
Nebraska	1578385	1287	288	0	18.0
Nevada	1201833	1563	297	1	30.5
New Hampshire	1109252	863	175	0	16.1
New Jersey	7730188	8489	1,079	1	14.0
New Mexico	1515069	2,016	538	1	35.6
New York	17990455	22568	2,345	1	13.1
North Carolina	6628637	8344	1669	1	25.7
North Dakota	638800	506	121	0	18.1
Ohio	10847115	11812	1,755	1	16.2
Oklahoma	3145585	3106	649	1	20.1
Oregon	2842321	2144	605	1	26.2
Pennsylvania	11881643	12270	2,028	1	16.9
Rhode Island	1003464	927	145	0	14.6
South Carolina	3486703	5107	1,012	1	29.2
South Dakota	696004	557	160	0	22.4
Tennessee	4877185	6152	1,404	1	28.7
Texas	16986510	21936	3491	1	20.7
Utah	1722850	2074	313	1	18.5
Vermont	562785	436	129	0	23.1
Virginia	6187358	7158	1,093	1	18.2
Washington	4866692	4177	817	1	17.6
West Virginia	1793477	1599	491	0	26.2
Wisconsin	1891769	4273	846	1	17.5
Wyoming	453588	515	159	1	33.1

Pop.(1990) = state's population as determined by the 1990 Census.
Low birth weights = number of births with child weighing under 2500g. From *Kids Count Data Book* produced by the Center for the Study of Social Policy.
Deaths in fatal accidents = number of fatalities in motor vehicle accidents in 1988

Deaths per 100000 people in fatal accidents = number of fatalities per 100000 people in the state in motor vehicle accidents in 1988
Belt laws-1990 = 0 if no seat belt law, 1 if seat belt law.
The last three variables are from "Do front-seat belt laws put rear-seat passengers at risk?"
Garbacz, C. Population Research and Policy Review, vol. 11: 157–168.

Is there a relationship between the number of motor vehicle accident deaths in a state and the presence of a seat belt law in the state? One way we might explore this relationship is to represent the absence or presence of a seat belt law by a code. Use the number 0 to mean the state has no seat belt law and 1 to mean the state has a seat belt law. While a variable indicating the absence or presence of a seat belt law is essentially a categorical variable, the code we have suggested reflects the following ordering. States with a seat belt law are more serious about seat belt use than states without a law. With this ordering in mind, we can talk about positive and negative association between seat belt laws and motor vehicle deaths. Positive association would mean that the states with seat belt laws have more motor vehicle deaths than states with no law. Negative association would mean that states with seat belt laws have fewer deaths than states with no law. If seat belt laws prevent deaths, we might expect a negative association between seat belt laws and motor vehicle deaths. To explore this, we use presence of a seat belt law as the explanatory variable and number of motor vehicle accident deaths in a state as the response variable. Notice that the explanatory variable is a state's law in 1990, while the number of traffic fatalities are from 1988. We were unable to obtain data on motor vehicle accident fatalities for 1990. We must assume that the 1988 fatalities are similar to those in 1990, or the laws in 1990 are similar to those in effect in 1988. As you will see, this discrepancy may not be serious. A scatterplot reveals the following.

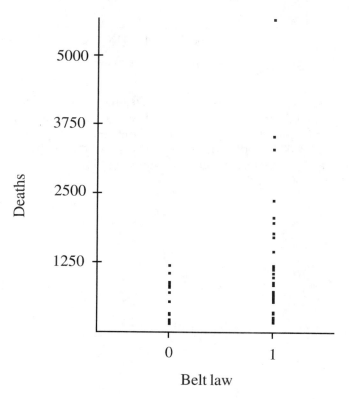

Surprisingly, we observe a positive association! Does this suggest that wearing seat belts may actually increase motor vehicle deaths? Perhaps people are trapped by faulty seat belts in a burning car and die because they are unable to escape the vehicle (one of the author's relatives believes this and dislikes using a seat belt). Perhaps people wearing seat belts perceive themselves to be safer and therefore drive more recklessly.

The population of the state may be a lurking variable. More populous states will have more fatal motor vehicle accidents simply because there are more people to get into accidents. If you look at the data you will notice that the states with the largest populations have seat belt laws. This may be in response to the higher number of fatalities in those states. If we make a scatterplot of a state's population (the explanatory variable) versus the number of motor vehicle accident fatalities (the response), and use different plotting symbols for the absence or presence of a law we can see the effect of the lurking variable. The symbol o represents states with law 0 (no law), | represents states with a law.

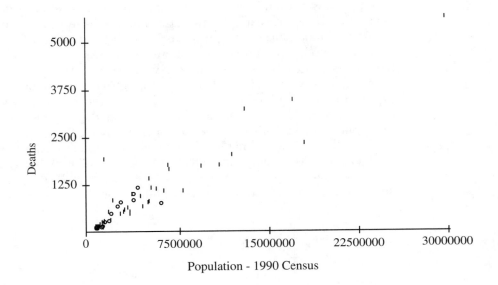

This suggests that a better way to investigate the relationship between seat belt laws and fatal motor vehicle accidents is to look at the fatality rate, i.e., the variable number of fatal accidents per person in the state. One of the variables in the data, deaths per 100000 people, provides a measure of the fatality rate. If we use this new variable as the response, we get the following plot.

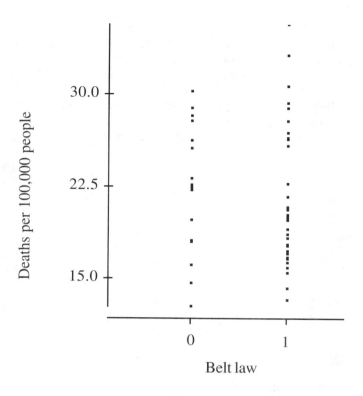

Now there does not appear to be much of a relationship between seat belt laws and fatal accidents, especially if one ignores the two states with the highest death rates. Adjusting for the population of the state has removed the association. Note that we have not proved that seat belt laws and motor vehicle deaths are not associated. They are associated. We have provided a plausible explanation for the observed association. This explanation undermines the claim that seat belt laws cause more motor vehicle deaths. Further examination of the data may reveal new associations. For example, it seems reasonable that number of motor vehicle deaths should be associated with the amount of driving that occurs in a state. This is not only due to the number of people in the state, but also involves the number of people that drive through the state. Perhaps adjusting for the number of miles of roads in a state would take into account the amount of driving that occurs in the state and would reveal new associations.

There are several lessons to be learned. First, any association you observe should be subject to careful scrutiny. Be on the alert for lurking variables. Be cautious before you jump to conclusions, especially if you intend to base an important decision on the observed association. It would be incorrect to repeal all seat belt laws on the basis of the initial positive association we observed between the severity of the law and number of motor vehicle deaths. Second, an observed association should not be interpreted as evidence of cause and effect without additional argument. In Chapter 3 we examine methods that allow us to make a case for cause and effect relationships. Third, be skeptical of reports of causal relationships when you are not given the details of how the conclusion was reached. In our case study, it would be easy to report our initial findings as evidence that seat belt laws should be repealed. We could write a very entertaining newspaper article about the positive association we observed. People reading our article (or hearing about it second hand) might be tempted to conclude that there is scientific evidence that wearing a seat belt is actually dangerous.

A final lesson involves the very search for association. In data sets with many variables, it is almost inevitable that one will find a reasonably strong association between some pair of variables. Observed associations that are the result of simply hunting around for an association among a large collection of variables should be handled with extreme caution. To illustrate this, we examined a large collection of variables for the 50 states to see if any were associated with traffic deaths. We found that the number of babies with low birth weights (births with the child weighing under 2500g) was strongly associated with traffic deaths. If we treat the number of babies with low birth weights as the explanatory variable and the number of traffic deaths as the response, a scatterplot of these variables is given below.

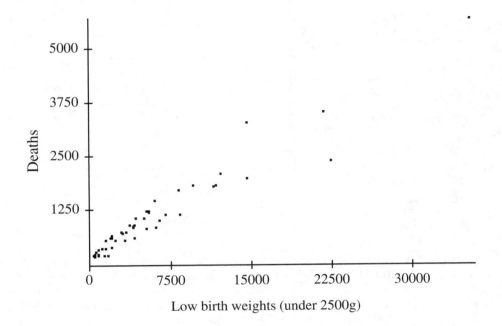

Low birth weights (under 2500g)

This is a very strong association. The correlation is 0.966. It would be tempting to conclude that something remarkable is occurring. Perhaps parents of babies with low birth weights are so concerned about the health of their child that they are not paying attention to their driving and are getting in more accidents! Once again, population size is a lurking variable here. We conjecture that if we knew the annual consumption of Twinkies in each state, that we would find a positive association between traffic deaths and Twinkie consumption (why?). Perhaps we should regulate Twinkies.

CHAPTER 3

PRODUCING DATA

CHAPTER OVERVIEW

Have you ever been just about to eat when someone conducting a survey calls and asks a bunch of questions? Did you answer the questions honestly? Did you ever wonder how the data were going to be used or why you were chosen for the survey? Surveys like this are just one way people collect data. Chapter 3 is all about the process of collecting high quality data.

In the first chapters, we have had data given to us, and we have assumed that they are good. By good, we mean useful in answering the questions that we have. This certainly isn't the case in the world outside of textbooks. Data are produced in many ways and not all of the methods used will generate usable data. There are two main ways of producing statistically useful data: designed observational studies and designed experiments. Both are based on the concept of sampling and the use of randomization. Good samples will represent the population we are interested in, and randomization helps ensure that. We shall see the good ways, simple random samples for example, and bad ways, voluntary response surveys for example, of obtaining samples and the importance of randomization.

SECTION 3.1

SECTION OVERVIEW

If a car manufacturer needs to test the strength of the side panels, one way is to test all the cars and see at what speeds they can withstand impact. There is just one problem with this method: no one will buy the damaged cars! Obviously testing all the cars just is not reasonable. This is one reason why we take **samples**, i.e., only examine a portion of the objects of interest. How the sample is chosen, that is, the **design**, has a large impact on the usefulness of the data. A useful sample will be representative of the **population** and will help answer our questions. "Good" methods of collecting a sample include the following:

probability samples

simple random samples (SRS)

stratified random samples

multistage samples

All these sampling methods involve some aspect of randomness through the use of a formal chance mechanism. Random selection is just one precaution that a person can take to reduce **bias,** the systematic favoring of a certain outcome. Samples we select using our own judgment, because they are convenient, or "without forethought" (mistaking this for randomness) are usually biased. It is in our nature to be biased in some way, even unconsciously. This is why we use computers or a tool like a **table of random digits** to help us select a sample.

Poor sample designs (or "non-designs") include the **voluntary response sample** and the **convenience sample**. Both of these methods rely on personal decision for the selection of the sample; this is generally a guarantee of bias in the selection of the sample. Even when trying very hard to be unbiased, the sample will generally not be truly representative of the population if a person does the choosing.

Other kinds of bias to be on the lookout for include **nonresponse bias**, bias in the **wording of questions**, and **undercoverage**. These types of bias can occur even in a randomly chosen sample and we need to try to reduce their impact as much as possible.

Remember, when selecting a sample, bigger is better.

KEY CONCEPTS

Population versus sample

The population is the whole group you are interested in learning something about. The sample is the part of the whole that you actually examine in order to learn about the whole.

How to choose a simple random sample

There are many ways to choose a simple random sample. A simple random sample, or SRS, of size n is a collection of n individuals chosen from the population in a manner so that each possible set of n individuals has an equal chance of being selected. In practice, simple methods such as drawing names (or slips of paper with the names written on them) from a hat is one way of getting an SRS. All the "names in the hat" are the population. To choose the sample, all we have to do then is mix up the "names" and pull out some. In reality the population may be all the voters in a city. We then need a list of the voters and usually a computer is put to use in pulling the names from the hat. For small samples we can use slips of paper in a hat or draw marbles from a box or almost any other chance mechanism.

Using a table of random digits is an excellent substitute for a computer selection. The text explains how to use these tables very well, but the method can be summed up in two steps.

1. Give every individual in the population its own numerical label. All labels need to have the same number of digits.

2. Starting anywhere in the table (usually a spot selected at random), read off labels until you have as many selected as needed for the sample.

How to choose a stratified random sample

1. Divide the population into strata. Form the strata based on some known characteristic of each individual. Common choices are location and age. Individuals in a particular stratum should be more like one another than those in the other strata.

2. Choose an SRS from each strata separately.

3. Combine the chosen individuals to form the whole sample.

Types of bias

Bias is the biggest problem in survey sampling. Bias in selecting the sample is controlled by randomization but we still have other problems.

Undercoverage is when some group in the population is given either no chance or a much smaller chance than other groups to be in the sample.

Nonresponse bias occurs when individuals who are selected do not participate or cannot be contacted.

Response bias is when individuals do participate but are not responding truthfully or accurately due to the way the question is worded, the presence of an observer, fear of a negative reaction from the interviewer, or any other such source. This need not be an intentional reaction on the part of the individual responding.

SOLUTIONS TO SELECTED TEXT EXERCISES

Exercise 3.3

This is a voluntary response sample. A person must see the advertisement in USA Today or watch the television show to be aware of the poll. People must pay to participate. Finally, participants are likely to be those that feel strongly about handgun control. All these factors make it impossible to assess whether those that vote are representative of the population. Experience suggests that such polls are almost certainly biased, usually in the favor of those opposing the question (in this case, those opposing handgun control). An added concern is that a person may vote more than once by calling repeatedly.

Exercise 3.5

There are 28 names in the list. We require two digits to label each name and so label them from 01 to 28. The result is given below.

01 Agarwal	08 Dewald	15 Huang	22 Puri
02 Anderson	09 Fernandez	16 Kim	23 Richards
03 Baxter	10 Fleming	17 Liao	24 Rodriguez
04 Bonds	11 Gates	18 Mourning	25 Santiago
05 Bowman	12 Goel	19 Naber	26 Shen
06 Castillo	13 Gomez	20 Peters	27 Vega
07 Cross	14 Hernandez	21 Pliego	28 Wang

We begin reading Table B at line 139, two digits at a time. The first 6 two-digit numbers corresponding to labels in the above list constitute our sample. We reproduce line 139 below (continuing on to additional lines as necessary) with two-digit numbers separated by vertical lines and the selected two-digit numbers underlined. Any repeats of previously selected numbers are skipped.

line 139: 55|58|8 9|94|04| 70|70|8 4|10|98| 43|56|3 5|69|34| 48|39|4 5|17|19|

line 140: 12|97|5 1|32|58| 13|048 ...

The sample is

04 = Bonds, 10 = Fleming, 17 = Liao, 19 = Naber, 12 = Goel, 13 = Gomez.

Exercise 3.9

We select a sample of size 25 (5% of 500) from accounts between $1000 and $50,000 and a sample of size 44 (1% of 4400) from accounts under $1000. We would label the 500 accounts between $1000 and $50,000 from 001 to 500, using three digits for each label. We would label the 4400 accounts under $1000 from 0001 to 4400, using four digits for each label.

To select the first 5 accounts from each of the strata, starting at line 115 in Table B, we begin reading in groups of three digits. Once we have found 5 three-digit numbers between 001 and 500, we begin reading the table in groups of four digits. When we have found 5 four digit numbers between 0001 and 4400, we are done. Line 115 is reproduced below (continuing onto additional lines as necessary), groups of digits divided by vertical lines, and selected digits underlined. We skip repeats of previously selected numbers.

line 115: 610|41 7|768|4 94|322| 247|09 7|3698| 1452|6 318|93 32|592

line 116: 1|4459| 2605|6 314|24 80|371 6|5103 ...

The sample from accounts between $1000 and $50,000 are those with labels 417, 494, 322, 247, and 097. The sample from accounts below $1000 is 3698, 1452, 2605, 2480, and 3716.

Exercise 3.11

We believe the July 1 to August 31 period had the higher rate of no answers. July 1 to August 31 is when most people take vacation and are out of town. This would lead to a higher rate of "ring-no-answer".

Nonresponse may lead to undercoverage if the nonrespondents differ in some systematic way from respondents. This is often the case. The higher the rate of nonresponse, the greater the undercoverage, and the less reliable the results. In the question, those most likely to be "ring-no-answers" between July 1 to August 31 are those who can afford to be on vacation away from home. The wealthier an individual or family, the longer their vacation is likely to be, and the more likely they are to be a "ring-no-answer."

Exercise 3.17

Call-in-polls are likely to be biased because they are voluntary response. Experience suggests that those most likely to take the time to participate (and pay the cost of participation in the case of "900" number phone-in-polls) are those that feel strongly against the issue being discussed and wish to voice their dissatisfaction. Another source of bias is that call-in-polls are generally conducted by radio or TV stations. In many instances these stations appeal only to a particular segment of the population (for example, consider the audiences of classical music radio stations, country music stations, and pop music stations). Listeners are not representative of the entire population. Only listeners are aware of the poll and hence likely to participate.

In the Reagan-Carter debate, we suspect that those that had strong, negative feelings about the then current administration (Carter's) were most likely to call. These would tend to be Republicans. At the time, inflation was very high (affecting business profits and investments) and Americans were being held captive by Iran. Both issues created negative feelings about the Carter administration, particularly among the upper-middle and upper classes. Traditionally these groups tend to have a high proportion of Republican voters.

Exercise 3.19

We label the 25 containers from 01 to 25, using a two digit label for each. The list with labels is given below.

01 A1096	06 A1097	11 A1098	16 A1101	21 A1108
02 A1112	07 A1113	12 A1117	17 A2109	22 A2211
03 A2220	08 B0986	13 B1011	18 B1096	23 B1101
04 B1102	09 B1103	14 B1110	19 B1119	24 B1137
05 B1189	10 B1223	15 B1277	20 B1286	25 B1299

Line 111 is reproduced below (continuing onto additional lines as necessary), groups of digits divided by vertical lines, and selected digits underlined. We skip repeats of digits previously selected.

line 111: 81 | 48 | 6 6 | 94 | 87 | 60 | 51 | 3 0 | 92 | 97 | 00 | 41 | 2 7 | 12 | 38 | 27 | 64 | 9 3 | 99 | 50 |

line 112: 59 | 63 | 6 8 | 88 | 04 | 04 | 63 | 4 7 | 11 | 97 ...

Our simple random sample of three bottles is 12 = A1117, 04 = B1102, and 11 = A1098.

Exercise 3.23

This is not a simple random sample. In a simple random sample of size 250, every possible group of 250 engineers must have the same chance of being the sample actually selected. In this case, the sample selected must consist of exactly 200 male and 50 female engineers. Samples consisting entirely of 250 males or 250 females have no chance of being selected. Even though a sampling procedure gives every individual the same chance of being selected, it need not be simple random sampling.

SECTION 3.2

SECTION OVERVIEW

In Section 3.1 we covered good and bad techniques used in designing a sample survey. In this section we learn about the other side of collecting statistically useful information: **experiments**. In some situations, such as when we wish to observe the current state of a population, the sample survey is best. Many times we will wish to study the effect of a specific treatment or type of intervention on a population. In these cases the well-designed experiment will give us much more information than even a perfectly designed **observational study**. For example, imagine yourself making microwave popcorn. You want to see what length of time is best for popping in your microwave oven. You design an experiment to test this. Your **factors** will be the different lengths of time. You may choose five different times; these are the **levels** of your factor. The different bags of popcorn are the **experimental units**. A simple experiment based on this idea will tell us much more than just picking any time and hoping not to burn the popcorn.

This section has three main points.

1. Compare treatments using a **control group** and treatment groups.

2. Randomization!

3. **Replication**. Repeat the experiment.

These principles form the foundation of a well-designed experiment.

Of course there are the usual problems with bias and possible confounding. Another problem to watch out for is the lurking variable. Some problems unique to experimental designs are the **placebo effect**, lack of **blinding**, and **lack of realism**.

In the last and optional section we look at some variations on the basic experimental designs. **Block designs** create groups of similar experimental units and are similar to strata in stratified sampling. The **matched pairs design** is one of the simplest sorts of block designs.

KEY CONCEPTS

Three principles of experimental design

1. **Compare treatments**. Looking at one treatment in isolation doesn't really tell us anything. The placebo effect may be the cause. However, comparing groups lets us see what happens when all the conditions are the same except for the experimental treatments. If an untreated group is used, it is called the control group. Comparison, possibly using a control group, lets us control for lurking variables. Any lurking variables will be present in all groups and hence differences will be due to treatments, not lurking variables.

2. **Randomization**. Using the ideas we learned in Section 3.1, place the experimental units in the groups using random assignment. Completely randomized is best. Make groups equal sized. This usually makes it easier to analyze results.

3. **Replication**. Repeat the treatments as many times as possible. A group of ten units will show any effects of the treatments much more clearly than groups of three. Using more units helps us separate the treatment effects from random effects caused by chance by reducing the chance variation in the results.

Why bother with a designed experiment?

Comparing treatments makes sure that outside influences are the same on all groups. Randomization will tend to give us similar groups of experimental units. This ensures that the results of the experiment are due to the treatments and not chance. Experiments can show us cause and effect!

SOLUTIONS TO SELECTED TEXT EXERCISES

Exercise 3.27

(a) The experimental units are the pairs of pieces of package liner. These are the objects to which the treatments are applied.

(b) The factor (explanatory variable) is the temperature of the jaws when the pairs of pieces of package liner are sealed. It has four levels, the temperatures of $250°$ F, $275°$ F, $300°$ F, and $325°$ F.

(c) The response variable is the force needed to peel each seal. This is what is measured on the units after treatment.

Exercise 3.29

(a) The experimental units in this experiment are the batches of the product that will be processed. The response is the yield of the process.

(b) There are two factors in this experiment. They are temperature and stirring rate. Since two temperatures are used at each of three stirring rates, there are a total of six treatments. Here is a diagram indicating the layout of the treatments.

		FACTOR B Stirring Rate		
		60 rpm	90 rpm	120 rpm
FACTOR A	$50°$ C	1	2	3
Temperature	$60°$ C	4	5	6

(c) Since two batches of the product will be processed at each combination of temperature and stirring rate (at each of the six treatments), a total of $2 \times 6 = 12$ batches (experimental units) are needed.

Exercise 3.31

Here is a diagram representing the package liner experiment of Exercise 3.27.

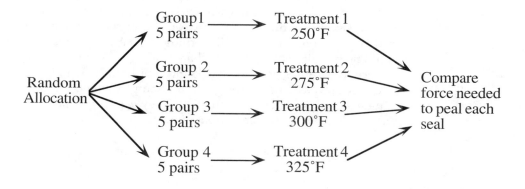

Exercise 3.33

Label the twenty pairs of pieces of package liners from 01 to 20, using two digits for each label. Read line 120 of Table B in groups of two digits. The liners corresponding to the first five two digit numbers between 01 and 20 selected get treatment 1. The next five get treatment 2, the next treatment 3, and the remaining five treatment 4. Line 120 is reproduced below (continuing onto additional lines as necessary), groups of digits divided by vertical lines, and selected digits underlined. We skip repeats of digits previously selected. Note that after 15 two-digit numbers between 01 and 20 are selected we are done, since they determine which liners get treatments 1, 2, and 3 (and, of course, the remainder get treatment 4).

```
line 120: 35|47|6 5|59|72| 39|42|1 6|58|50| 04|26|6 3|54|35| 43|74|2
          1|19|37|
line 121: 71|48|7 0|99|84| 29|07|7 1|48|63| 61|68|3 4|70|52| 62|22|4
          5|10|25|
line 122: 13|87|3 8|15|98| 95|05|2 9|09|08| 73|59|2 7|51|86| 87|13|6
          9|57|61|
line 123: 54|58|0 8|15|07| 27|10|2 5|60|27| 55|89|2 3|30|63| 41|84|2
          8|18|68|
line 124: 71|03|5 0|90|01| 43|36|7 4|94|97| 72|71|9 9|67|58| 27|61|1
          9|15|96|
line 125: 96|74|6 1|21|49| 37|82|3 7|18|68| 18|44|2 3|51|19| 62|10|3
          3|92|44|
line 126: 96|92|7 1|99|31| 36|80|9 7|41|92| 77|56|7 8|87|41| 48|40|9
          4|19|03|
line 127: 43|90|9 9|94|77| 25|33|0 6|43|59| 40|08|5 1|69|25| 85|11|7
          3|60|71|
```

Thus we have the following assignments.

Treatment 1 = those pairs labeled 16, 04, 19, 07, and 10.

Treatment 2 = those pairs labeled 13, 15, 05, 09, and 08.

Treatment 3 = those pairs labeled 18, 03, 01, 06, and 11.

Treatment 4 = those pairs remaining, which are those labeled 02, 12, 14, 17, and 20.

Exercise 3.37

"A significant difference" between the sexes means that the observed difference in earnings between men and women in the sample was too large to attribute plausibly to chance. The difference in earnings observed in the sample is therefore likely to indicate a real difference in earnings between the sexes in the entire population.

"No significant difference" between earnings of black and white students means that any observed difference in earnings in the sample was sufficiently small that it could be plausibly attributed to chance. In other words, suppose there is no real difference in earnings between black and white students in the population. Small differences, such as those observed in the sample, would not be unusual in additional similar samples from this population. They can simply be explained by the fact that no sample truly reflects the population exactly.

Exercise 3.39

(a) If all patients received the drug, we would have no way of knowing if any observed effect is due to the drug or to the placebo effect (favorable response to any treatment, even a dummy treatment). The placebo effect would be a lurking variable in such an experiment.

(b) Here is a diagram outlining the design of the experiment.

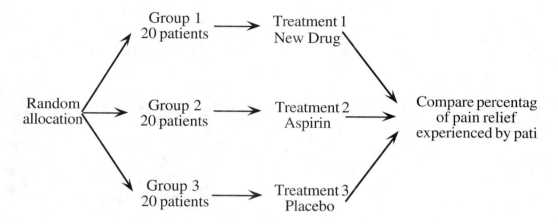

It is probably a good idea to run this experiment double blind (see answers to (c) and (d) below.

(c) If patients know which treatment they are receiving, this may effect their response. Those receiving the placebo will know they are not getting any real treatment and are not likely to notice any pain relief. Those receiving aspirin are likely to respond in a manner consistent with past experience with taking aspirin. Those receiving the new treatment may respond in a manner consistent with their expectations about this treatment. If, for example, they have heard that the new treatment is a powerful pain reliever they may respond positively due to a kind of placebo effect. Knowledge (and expectations) of the treatment they receive acts as a lurking variable in this case.

(d) Since the patient is required to report their (subjective) opinion concerning the percentage of pain relief experienced, one must be careful. The physician recording the response could conceivably affect a patient's response by prompting for a particular answer (for example if the physician expects the new drug to be very effective, the physician may prompt for a higher percentage than the patient initially suggests, particularly if the patient gives an uncertain response) or by tone of voice (for example, surprise if a patient on placebo indicates substantial pain relief). To avoid such sources of hidden bias, it is probably best to err on the safe side and require the experiment to be double blind.

Exercise 3.41

(a) Here is a diagram outlining the design of the experiment.

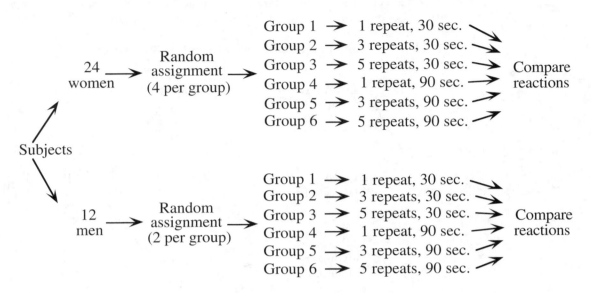

(b) We begin by selecting the women. We label these 01 to 24. Read line 140 of Table B in groups of two digits. The women corresponding to the first four two-digit numbers between 01 and 24 selected get treatment 1. The next four get treatment 2, the next treatment 3, then 4, then 5, and the remaining four treatment 6. Line 140 is reproduced below (continuing onto additional lines as necessary), groups of digits divided by vertical lines, and selected digits underlined. We skip repeats of digits previously selected. Note that after 20 two-digit numbers between 01 and 24 are selected we are done, since they determine which liners get treatments 1, 2, 3, 4, and 5 (and, of course, the remainder get treatment 6).

line 140: <u>12</u>|97|5 1|32|58| <u>13</u>|<u>04</u>|8 4|51|44| 72|32|<u>1 8</u>|<u>19</u>|40| 00|36|0 0|<u>24</u>|28|

line 141: 96|76|7 3|59|64| <u>23</u>|82|2 9|60|12| 94|59|<u>1 6</u>|51|94| 50|84|2 5|33|72|

line 142: 72|82|9 5|<u>02</u>|32| 97|89|2 6|34|<u>08</u>| 77|91|9 4|45|75| 24|87|0 0|41|78|

line 143: 88|56|5 4|26|28| <u>17</u>|79|7 4|93|76| 61|76|<u>2 1</u>|69|53| 88|60|4 1|27|24|

line 144: 62|96|4 8|81|45| 83|08|3 6|94|53| 46|<u>10</u>|9 5|95|<u>05</u>| 69|68|0 0|<u>09</u>|00|

line 145: 19|68|7 1|26|33| 57|85|7 9|58|<u>06</u>| 09|93|1 0|21|50| 43|16|3 5|86|36|

line 146: 37|60|9 5|90|57| 66|96|7 8|34|<u>01</u>| 60|70|5 0|23|84| 90|59|7 9|36|00|

line 147: 54|97|3 8|62|78| 88|73|7 7|43|51| 47|50|0 8|45|52| 19|90|9 6|71|81|

line 148: 00|69|4 0|59|77| 19|66|4 6|54|41| <u>20</u>|90|3 6|23|71| <u>22</u>|72|5 5|33|40|

line 149: 71|54|6 0|52|33| 53|94|6 6|87|43| 72|46|0 2|76|01| 45|40|3 8|86|92|

line 150: <u>07</u>| ...

Now we select the men. We label them 01 to 12. We begin reading Table B where we left off in line 150. We read in two-digit groups. The first two two-digit numbers selected get treatment 1, the next two treatment 2, etc. Once we have assigned men to the first five treatments, the remaining two men get treatment 6. As before, we skip any two-digit numbers assigned to men already selected as we read through the table. When we finish line 150, we start at the top of the table with line 101.

line 150: —51|1 8|89|15| 41|26|7 1|68|53| 84|56|9 7|93|67| 32|33|7 0|33|16|

line 101: 19|22|3 9|50|34| <u>05</u>|75|6 2|87|13| 96|40|9 1|25|31| 42|54|4 8|28|53|

line 102: 73|67|6 4|71|50| 99|40|0 0|19|27| 27|75|4 4|26|48| 82|42|5 3|62|90|

line 103: 45|46|7 7|17|<u>09</u>| 77|55|8 0|00|95| 32|86|3 2|94|85| 82|22|6 9|00|56|

line 104: 52|71|1 3|88|89| 93|<u>07</u>|4 6|<u>02</u>|27| 40|<u>01</u>|1 8|58|48| 49|76|7 5|25|73|

line 105: 95|59|2 9|40|07| 69|97|1 9|14|81| 60|77|9 5|37|91| 17|29|7 5|93|35|

line 106: 68|41|7 3|50|13| 15|52|9 7|27|65| 85|<u>08</u>|9 5|70|67| 50|21|1 4|74|87|

line 107: 82|73|9 5|78|90| 20|80|7 4|75|<u>11</u>| 81|67|6 5|53|00| 94|38|3 1|48|93|

line 108: 60|94|0 7|20|24| 17|86|8 2|49|43| 61|79|0 9|<u>06</u>|56| 87|96|4 1|88|83|

line 109: 36|00|9 1|93|65| 15|41|2 3|96|38| 85|45|3 4|68|16| 83|48|5 4|19|79|

line 110: 38|44|8 4|87|89| 18|33|8 2|46|97| 39|36|4 4|20|06| 76|68|8 0|87|08|

line 111: 81|48|6 6|94|87| 60|51|3 0|92|97| 00|41|2 7|<u>12</u>|38| 27|64|9 3|99|50|

line 112: 59|63|6 8|88|<u>04</u>| ...

The assignments to treatment are thus

TREATMENT

1	2	3	4	5	6
woman 12	woman 19	woman 02	woman 10	woman 01	woman 03
woman 13	woman 24	woman 08	woman 05	woman 20	woman 11
woman 04	woman 23	woman 17	woman 09	woman 22	woman 14
woman 18	woman 16	woman 21	woman 06	woman 07	woman 15
man 05	man 07	man 01	man 11	man 12	man 03
man 09	man 02	man 08	man 06	man 04	man 10

Note that we had to read a considerable portion of Table B before completing the random assignments. This is likely to happen when the numerical labels we use (in this case two-digit numbers between 01 and 24 or 01 and 12) are a small fraction of all possible labels (in this case all the two-digit numbers 00 to 99). Computer software can be used in practice to automate the process of random selection and this will eliminate much of the tedium of reading many lines of a random digits table such as Table B.

Exercise 3.43

In this exercise, there are two treatments. These are whether or not one uses software that highlights trends. The twenty students are the subjects. The response is the amount of money made by a student. We describe two designs.

DESIGN 1

Divide the students into two groups of 10 by random allocation. One group uses the software that highlights the trends. The other group does not use the software. Compare the amount of money made by each group. A diagram describing the design is given below.

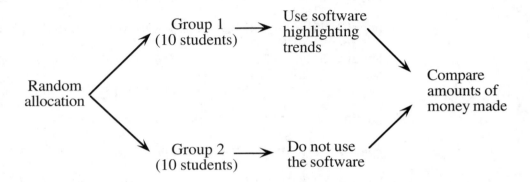

If we use this design, we need to assign the 20 students to the two treatments. We label the students from 01 to 20 and begin at line 125 of Table B. We read the table in pairs of digits. The first ten digits between 01 and 20 we find are assigned to group 1, the remainder to group 2. Line 125 is reproduced below (continuing onto additional lines as necessary), groups of digits divided by vertical lines, and selected digits underlined. We skip repeats of digits previously selected.

    line 125: 96|74|6 1|21|49| 37|82|3 7|18|68| 18|44|2 3|51|19| 62|10|3
              3|92|44|
    line 126: 96|92|7 1|99|31| 36|80|9 7|41|92| 77|56|7 8|87|41| 48|40|9
              4|19|03|
    line 127: 43|90|9 9|94|77| 25|33|0 6|43|59| 40|08|5 1|69|25| 85|11|7
              3|60|71|
    line 128: 15|68|9 1|42|27| 06|56|5 1|43|74| 13|35|2 4|93|67| 81|98|2
              8|72|09|

The results are

        Group 1: students 18, 19, 10, 03, 06, 08, 11, 15, 13, 09
        Group 2: students 01, 02, 04, 05, 07, 12, 14, 16, 17, 20

DESIGN 2

Divide the students into two groups of 10 by random allocation. One group first uses the software that highlights the trends for a given period of time and the total amount of money made by each student is recorded. These students then do not use the software for the same period of time and the total amount of money made by each student during this second period is recorded. The other group does not use the software during the first period, then switches to using the software during the second period. The total made by each student in each period is recorded. Compare the amount of money made during the period when students were using the software to the amount made when not using the software. A diagram describing the design is given below.

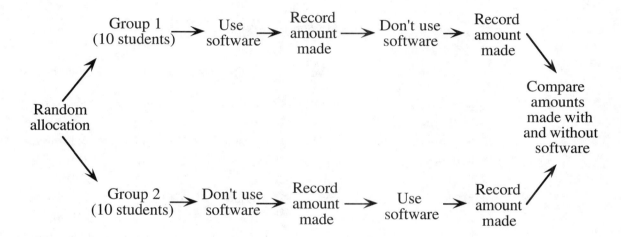

The randomization can be carried out just as in DESIGN 1 to assign students to groups. See DESIGN 1 for details.

An alternative that you might consider is to flip a coin (or select a random digit). If you get heads (or a number between 0 and 4), assign the student to group 1 (use the software first). If you get tails (or a number between 5 and 9), assign the student to group 2 (use the software second). If this method is used, there is no guarantee that you will get exactly 10 students in each group. For example, if we use line 130, reading one digit at a time, assigning student i to group 1 if the ith digit is between 0 and 4. If we find

line 130: 69051 64817 87174 09517...

Group 1: students 3, 5, 7, 9, 13, 15, 16, 19
Group 2: students 1, 2, 4, 6, 8, 10, 11, 12, 14, 17, 18, 20

which puts 8 students in group 1 and 12 in group 2. You might try balancing the groups using this method by stopping as soon as one group gets 10 students and put the remainder in the other group. If this strategy were followed we would have assigned students 18 and 20 to group 1 rather than group 2. Although it is not obvious, this method is not exactly like simple random sampling. It turns out that this method would tend to place students 19 and 20 together too often in the same group (if you think about it you can see that there may be a tendency to get 10 students in one of the groups by the time you reach student 19, so that students 19 and 20 get put together in the remaining group).

Exercise 3.51

(a) We use a placebo as a second treatment and run the experiment double blind. Here is a diagram outlining the design of this experiment.

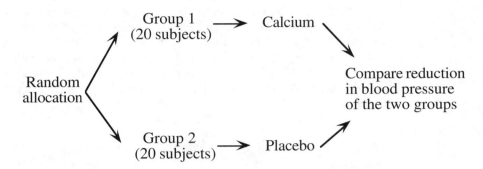

(b) We label the subjects from 01 to 40. The labels and names are given below.

01 Alomar	09 Denman	17 Han	25 Liang	33 Rosen
02 Asihiro	10 Durr	18 Howard	26 Maldonado	34 Solomon
03 Bennett	11 Edwards	19 Hruska	27 Marsden	35 Tompkins
04 Bikalis	12 Farouk	20 Imrani	28 Moore	36 Townsend
05 Chen	13 Fratianna	21 James	29 O'Brian	37 Tullock
06 Clemente	14 George	22 Kaplan	30 Ogle	38 Underwood
07 Cranston	15 Green	23 Krushchev	31 Plochman	39 Willis
08 Curtis	16 Guillen	24 Lawless	32 Rodriguez	40 Zhang

We begin at line 119 of Table B. We read the table in pairs of digits. The first twenty digits between 01 and 40 we find are assigned to group 1 (drug), the remainder to group 2 (placebo). Line 119 is reproduced below (continuing onto additional lines as necessary), groups of digits divided by vertical lines, and selected digits underlined. We skip repeats of digits previously selected.

line 119: 95|85|7 0|71|<u>18</u>| 87|66|4 9|<u>20</u>|99| 58|80|6 6|69|79| 98|62|4
     8|48|<u>26</u>|
line 120: <u>35</u>|47|6 5|59|72| <u>39</u>|42|<u>1 6</u>|58|50| <u>04</u>|26|6 3|54|35| 43|74|<u>2</u>
     <u>1</u>|<u>19</u>|<u>37</u>|
line 121: 71|48|7 0|99|84| <u>29</u>|<u>07</u>|7 1|48|63| 61|68|<u>3 4</u>|70|52| 62|<u>22</u>|4
     5|<u>10</u>|<u>25</u>|
line 122: <u>13</u>|87|<u>3 8</u>|<u>15</u>|98| 95|<u>05</u>| ...

Those receiving the drug are thus

18 = Howard	20 = Imrani	26 = Maldonado	35 = Tompkins
39 = Willis	16 = Guillen	04 = Bikalis	21 = James
19 = Hruska	37 = Tullock	29 = O'Brian	07 = Cranston
34 = Solomon	22 = Kaplan	10 = Durr	25 = Liang
13 = Fratianna	38 = Underwood	15 = Green	05 = Chen

Exercise 3.53

The placebo effect occurs when a patient with a *real* illness responds to a treatment, even if the treatment is a dummy medication, simply because it is a treatment. Thus people with real illnesses can show improvement due to the placebo effect. Response to a placebo need not mean that the patient had no problem to begin with.

SOLUTIONS TO SELECTED TEXT REVIEW EXERCISES

Exercise 3.57

(a) There will be 3478 students on the list. Since 3478 is a four-digit number, we might label them from 0001 to 3478, using four digits for each label.

(b) We begin at line 105 of Table B. We read the table in groups of four digits. The first 100 different four-digit numbers between 0001 and 3478 will be the labels of the students comprising our sample. Line 105 is reproduced below (continuing onto additional lines as necessary), groups of digits divided by vertical lines, and selected digits underlined. We skip repeats of digits previously selected. We only select the first five students.

line 105: 9559|2 940|07 69|971 9|1481| 6077|9 537|91 17|297 5|9335|
line 106: 6841|7 350|13 15|529 ...

The first five students in our sample are therefore the students labeled 2940, 0719, 1481, 2975, and 1315.

Exercise 3.61

(a) Here is a graphic displaying the design of the experiment.

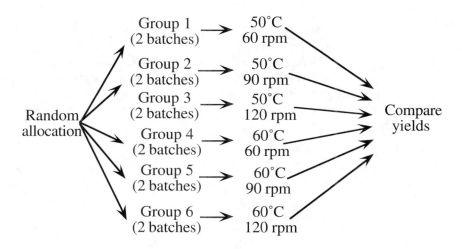

(b) Label the twelve batches from 01 to 12, using two digits for each label. Read line 129 of Table B in groups of two digits. The batches corresponding to the first two-digit numbers between 01 and 12 selected get treatment 1. The next two get treatment 2, the next treatment 3, the next 4, the next 5, and the remaining two treatment 6. The order in which the batches are selected also determines the order in which they will be processed. Line 128 is reproduced below (continuing onto additional lines as necessary), groups of digits divided by vertical lines, and selected digits underlined. We skip repeats of digits previously selected. Note that after 10 two-digit numbers between 01 and 12 are selected we are done, since they determine which liners get treatments 1, 2, 3, 4, and 5 (and, of course, the remainder get treatment 6).

line 128: 15|68|9 1|42|27| <u>06</u>|56|5 1|43|74| 13|35|2 4|93|67| 81|98|2
8|72|<u>09</u>
line 129: 36|75|9 5|89|84| 68|28|8 2|29|13| 18|63|8 5|43|<u>03</u>| 00|79|5
0|87|27|
line 130: 69|<u>05</u>|1 6|48|17| 87|17|4 0|95|17| 84|53|4 0|64|89| 87|20|1
9|72|45|
line 131: 05|00|7 1|66|32| 81|19|4 1|48|73| <u>04</u>|19|7 8|55|76| 45|19|5
9|65|65|
line 132: 68|73|2 5|52|59| 84|29|2 0|87|96| 43|16|5 9|37|39| 31|68|5
9|71|50|
line 133: 45|74|9 4|18|<u>07</u>| 65|56|1 3|33|<u>02</u>| 07|05|1 9|36|23| 18|13|2
0|95|47|
line 134: 27|81|6 7|84|16| 18|32|9 2|13|37| 35|21|3 3|77|41| 04|31|2
6|85|<u>08</u>|
line 135: 66|92|5 5|56|58| 39|<u>10</u>|0 7|84|58| <u>11</u>|20|6 1|98|76| 87|15|1
3|<u>12</u>|60|

The assignment is then

Treatment 1 = batches 06, 09
Treatment 2 = batches 03, 05
Treatment 3 = batches 04, 07
Treatment 4 = batches 02, 08
Treatment 5 = batches 10, 11
Treatment 6 = batches 01, 12 (I flipped a coin to see which of these two
was listed first)

The numbers are listed in the order in which they should be processed according to each treatment.

Note that in addition, we might consider using randomization to determine which treatment is run first, second, third, etc., up to which is run twelfth (remember, each treatment is observed twice). This would be important if the order in which batches are processed could affect the yield in any systematic way.

Exercise 3.65

"Statistically significant" means that the observed effect is too large to attribute plausibly to chance. In particular, if the score on the screening test and the score on an evaluation are not actually correlated, then it would be very unlikely to obtain a correlation as large as was observed.

CASE STUDY

The case study below is based on an article by Lafferty, M. B. (1993), "OSU scientists get a kick out of sports controversy", The Columbus Dispatch (November, 21, 1993), B7.

In the Autumn of 1993, Auburn University played Mississippi State University in football. Faced with a fourth down play deep in their own territory, Auburn set up to punt. Mississippi State looked forward to good field position following the punt. The football was snapped, kicked and eyed in disbelief as it sailed an estimated 71 yards through the air. Shocked, the Mississippi State coaches claimed the football must have been filled with helium in order to produce such a kick. The football was immediately seized by the officials and was later tested to see if it had been filled with helium. No helium was found, but the possible benefits of filling a football with a gas lighter than air would be kicked around both science and sports communities in the weeks to come. Do helium-filled balls travel further than footballs filled with air?

Two experiments were conducted by members of the media in Columbus, Ohio to investigate this question. The first experiment was carried out by a team from WBNS television. According to a report in the Columbus Dispatch (the local newspaper) by Lafferty (1993), the WBNS study was conducted using two different footballs, one of which was filled with helium while the other was filled with ordinary air. Each football was kicked four times with the wind and four times against the wind. The kicker was aware of which of the two footballs he was kicking. The WBNS study reported that a helium-filled football traveled an average of 10 yards farther with the wind and an average of 5 yards less against the wind than its air-filled counterpart.

At first glance, the results of this study seem striking. As one might expect, the lighter, helium-filled ball went much farther (10 yards) when the wind was at the kickers back, but did not perform so well into the wind. Is this evidence of the advantage of kicking a helium-filled football, at least when the wind is at your back?

Unfortunately, there are several flaws in this first study. First, the study was not blind. The kicker knew which football he was kicking. This knowledge could effect the way the kicker kicked the football, kicking the helium-filled football more smoothly than the air-filled ball when the wind was at his back, while lunging at the helium-filled ball when kicking into the wind. A smoother rhythm generally produces a longer kick. Lack of blindness confounds the results with a sort of "placebo effect".

A second flaw is the small sample size. Observed differences in small scale studies are often attributable to chance, especially if there is considerable variability in the individual results. Furthermore, a single outlier (one flub with the air-filled football or one spectacular kick with the helium-filled football) could account for the observed differences. To determine if there is considerable variability in the data or if there are outliers, we would like to see the actual data. Unfortunately, the data were not given. As we have repeatedly emphasized, it is difficult to evaluate the results of a study if you are not given the actual data.

Finally, we do not know if any randomization was used in the study. Was the order of kicks randomized? Was randomization used to determine which football was filled with air and which with helium? One would want to control for differences in the footballs (if any existed) perhaps by using several footballs, some filled with helium and some with air, the determination of which football was filled with which substance by some sort of randomization. Obviously one would also hope that the footballs were filled to similar pressures and, as far as possible, conditions were kept constant for all kicks. We are not told if this was the case in the WBNS study.

After hearing these results, skeptics from The Columbus Dispatch decided to conduct their own experiment with the help of a team of physicists and chemists from The Ohio State University. In this study, two identical footballs, one air-filled and one helium-filled, were used outdoors on a windless day at The Ohio State University's athletic complex. A casual observer was unable to find a difference in the pressure of the two footballs and it was noted that the actual weight difference between these equal volumes of air and helium is a few thousandths of a pound compared with the weight of the football itself which weighs nearly a pound (weights are reported below). The kicker was a novice punter and was not informed which football contained the helium (thus the study was blind). Each football was kicked 39 times (replication) and the two footballs were alternated with each kick. A pair of consecutive kicks of the two different balls thus serves as a sort of matched pair.

This study is better designed than the WBNS study. The actual data along with a summary of the results are given below.

ACTUAL DATA

Trial	Helium	Air	Helium-Air
1	25	25	0
2	16	23	-7
3	25	18	7
4	14	16	-2
5	23	35	-12
6	29	15	14
7	25	26	-1
8	26	24	2
9	22	24	-2
10	26	28	-2
11	12	25	-13
12	28	19	9
13	28	27	1
14	31	25	6
15	22	34	-12
16	29	26	3
17	23	20	3
18	26	22	4
19	35	33	2
20	24	29	-5
21	31	31	0
22	34	27	7
23	39	22	17
24	32	29	3
25	14	28	-14
26	28	29	-1
27	30	22	8
28	27	31	-4
29	33	25	8
30	11	20	-9
31	26	27	-1
32	32	26	6
33	30	28	2
34	29	32	-3
35	30	28	2
36	29	25	4
37	29	31	-2
38	30	28	2
39	26	28	-2

SUMMARY STATISTICS

	Air-filled	Helium-filled
Ball weight*	0.86 lb	0.82 lb
Mean distance	26.0 yards	26.4 yards
Median distance	26 yards	28 yards
Longest kick	35 yards	39 yards

*Lafferty (1993) attributes this difference to variation in football materials rather than the different gases used.

We note several things. First, the difference in the weight of the two balls is .04 lbs. This is probably not a difference a person could detect by feel and hence it is hard to imagine how such a small weight difference could cause a large difference in the length of kicks. Second, the helium-filled football yielded a longer mean and median distance and also a longer maximum. However, there is considerable variation in the distances kicked and the helium-filled ball produced the four shortest kicks. A better means of comparing the results might be to look at histograms of the kicks. These histograms are given below.

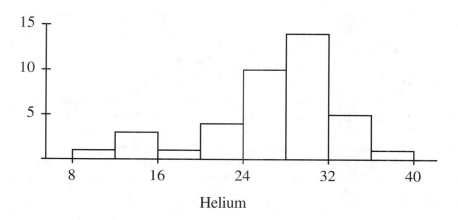

Helium

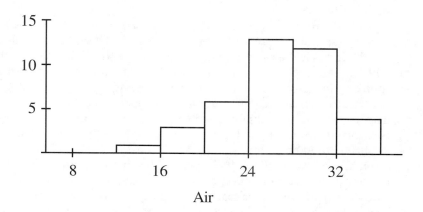

Air

We notice that the kicks for both the helium and air-filled footballs show a lot of variability, with greater variability with the helium-filled ball than the air-filled ball. The distributions of both are slightly skewed to the left but do have a rough bell shape. The center of the helium-filled football data seems a bit larger than that for the air-filled football, as indicated by the mean and median of the data. The difference is small and the variability in the data (especially the greater variability in the helium-filled football data) makes it hard to assert that there is any marked advantage to kicking the helium-filled football.

We mentioned that the pair of kicks comprising a given trial might be viewed as a matched pair. For such data, it is often valuable to examine the difference in the pair of values comprising the matched pair. The differences (helium-/air-filled results) are listed in the data and a histogram of the values is given below.

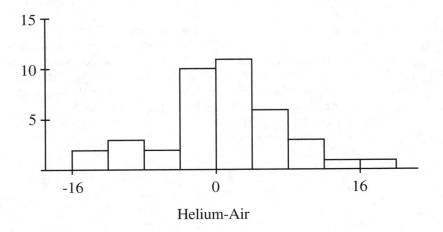

Helium-Air

This histogram is roughly symmetric and bell shaped. It's center is around 0, perhaps slightly above 0. There is considerable variation in the differences. Once again, it is hard to see that there is any marked advantage to kicking a football filled with helium versus one filled with air. At best there is weak evidence of a slight advantage for the helium-filled football. Certainly, these results do not substantiate the WBNS study, which seemed to suggest a much clearer advantage to the helium-filled football.

What do we conclude? First, there does not seem to be much evidence that a helium-filled football outperforms an air-filled football. The next time someone "booms" a long punt or field goal, and the announcers question whether the football is filled with helium, be skeptical.

Second, we might ask why the WBNS study found such a large difference in the performance of helium and air-filled footballs. One possibility is a sort of "placebo effect". If this is the case, all that matters is that the kicker believe the football is filled with helium (and believe such a football will fly farther than one filled with air). This could be investigated with another well-designed experiment (we don't think the WBNS study is sufficient to draw such conclusions because of its design flaws). If true, coaches don't actually need to fill footballs with helium. They simply need to convince their kickers that the football is filled with helium. This might be a strategy a coach should consider pursuing!

Finally, be wary of results of poorly designed studies. Certainly, do not base any important decision or action on such a study. As is the case here, a better designed study often fails to corroborate such results. Be on the alert for lack of randomization, lack of blindness, and small sample sizes. These are often characteristics of poorly designed studies.

CHAPTER 4

SAMPLING DISTRIBUTIONS AND PROBABILITY

CHAPTER OVERVIEW

So far in this book we have covered how to collect data (Chapter 3) and some basic ways of looking at these data (Chapters 1 and 2). The methods we explored in the first two chapters are called exploratory data analysis. These are great beginnings but rarely are these tools the end of the line. The more formal methods of analyzing data are called statistical inference. Statistical inference, a big topic, is the subject of the rest of the book. Inference is the tool we will use to draw concrete conclusions about data. Inference is most effective when used in conjunction with the exploratory data analysis tools we already know how to use.

Suppose we are interested in whether or not listening to classical music helps improve test scores. We can think of the difference between the two methods like this: in exploratory analysis we may notice a pattern in the test scores and it may seem like tests taken after listening to classical music are higher. All we can say is "There seems to be a pattern." If we use statistical inference we will be able to determine the strength of the evidence provided by the pattern. Could the pattern merely have happened by chance or is there strong evidence that listening to classical music increases test scores? We may even be able to say by how much the music increases the scores and if the increase is significantly different from the increase in scores due to, say, classic rock music. Inference is very useful!

The foundation of statistical inference is probability. Probability is key. In this chapter we will learn the language of probability. Chapter 5 will show us how to connect probability to inference. Some of the sections in the chapter are optional and not needed to understand the rest of the book. If the optional sections are covered in your class then the sections here may be of help. If you are not covering the optional material in class just ignore those sections in the Study Guide.

SECTION 4.1

SECTION OVERVIEW

This first section introduces **parameters** and **statistics.** A statistic is a number we calculate based on a sample from the population and has no unknown parameters. In the real world we rarely know the population parameters and we need statistics to estimate them. Good estimates are **unbiased**. We measure the usefulness of a statistic according to its distribution. The distribution of a statistic is called the **sampling distribution**. These distributions are described in the same way as the distributions we met earlier. We need

- a measure of center: the mean
- a measure of spread: standard deviation, variance
- a description of the shape of the distribution
- to look for outliers and patterns

Probability theory will give us the exact distribution of a statistic. One important item is that larger samples will give statistics that are better estimates of the parameter. Probability is the long-run proportion of times an outcome occurs. These probability rules apply to randomly selected samples but a sample that is not random may not give useful statistics.

KEY CONCEPTS

What is unbiased?

A statistic is an unbiased estimate of the population parameter if the mean of the statistic's distribution is equal to the mean of the parameter. So to check if a statistic is unbiased we need to know its mean and to compare it to the mean of the parameter the statistic is an estimate of. Probability is the tool we use to check these things.

> $\hat{p}$ is an unbiased estimate of p.

Sampling distributions

Just like the distributions we encountered in Chapter 1, we will use the same measures of center and spread to judge the value of a statistic. The spread of the sampling distribution tells us the variability of the statistic. Small spread means there will be less variability in the statistic that means the statistic is a good estimate. Larger samples produce statistics with smaller spread.

Rules of probability

1. Probabilities are numbers from 0 to 1. $P(A)$ means "the probability of A." If the probability is 0, the event will never occur. If the probability is 1, the event will always occur.

2. All the possible outcomes together have the probability of 1.

3. The probability of an event occurring is equal to one minus the chance of the event not occurring. In short hand this is written

$$P(A) = 1 - P(\text{not } A)$$

4. If two events have no outcomes in common, then the probability of one or the other occurring is the sum of their two probabilities.

➤ There are three snakes in a bag. You will only pull out one. Snake A is very small and has a 20% chance of being caught. The other two each have a 40% chance of being caught. The probability of catching snake A is 0.20. The probability of not catching snake A is $1 - 0.20 = 0.80$. The chance of catching snake A or B is $0.20 + 0.40 = 0.60$. The probability of catching any of the snakes is $0.20 + 0.40 + 0.40 = 1.0$.

SOLUTIONS TO SELECTED TEXT EXERCISES

Exercise 4.3

4.8% is a statistic since it describes the sample of 100 numbers dialed. 52% is a parameter since it describes the population of all Los Angeles residential phones.

Exercise 4.5

(a) We tossed a coin 20 times and got 8 heads. Thus $\hat{p} = 8/20 = 0.40$.

(b) Repeating this process 10 times we obtained 12, 9, 10, 7, 11, 5, 9, 10, 11, and 16 heads. Dividing each by 20 gives 0.60, 0.45, 0.50, 0.35, 0.55, 0.25, 0.45, 0.50, 0.55, and 0.80 for the values for $\hat{p}$. A histogram of these results is

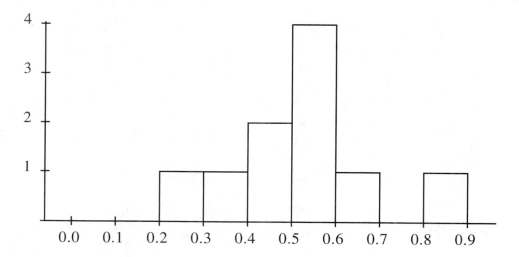

p

We see that the center is close to 0.5.

(c) We obtained 250 repetitions of twenty coin tosses. A histogram of the values of $\hat{p}$ is

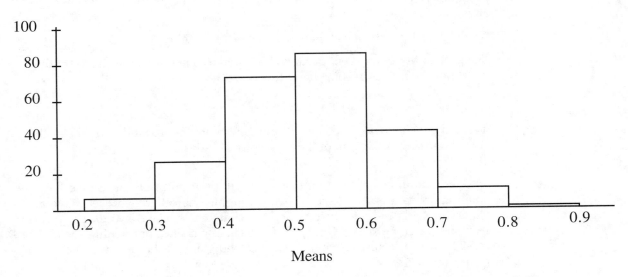

Means

The center is close to 0.5 and the shape is approximately normal.

Exercise 4.7

(a) Here is a table that shows how often each count occurs and the corresponding value of the sample proportion.

Count	$\hat{p}$	Frequency
9	0.045	1
10	0.050	0
11	0.055	0
12	0.060	0
13	0.065	3
14	0.070	2
15	0.075	5
16	0.080	11
17	0.085	12
18	0.090	12
19	0.095	9
20	0.100	7
21	0.105	5
22	0.110	6
23	0.115	7
24	0.120	10
25	0.125	4
26	0.130	1
27	0.135	2
28	0.140	2
29	0.145	0
30	0.150	1

A histogram of these values of $\hat{p}$ with bars of width 0.02 is

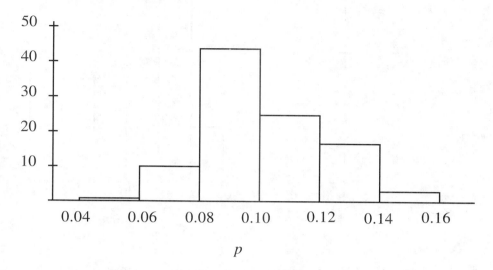

A histogram of these values of $\hat{p}$ with bars of width 0.01 is

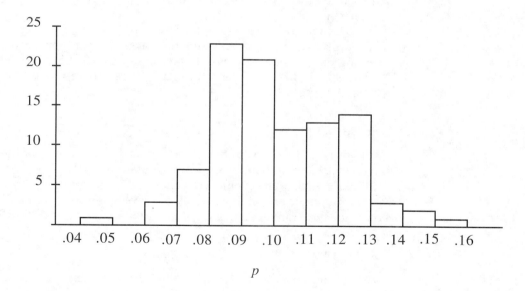

(b) The histogram with bars of width 0.20 looks approximately normal, perhaps it is slightly right skewed. The histogram with bars of width 0.1 looks a bit less normal and slightly left skewed.

(c) The mean of the 100 observations is 0.0981, which we calculated with statistical software. We mark the mean on the histogram with bar width 0.01.

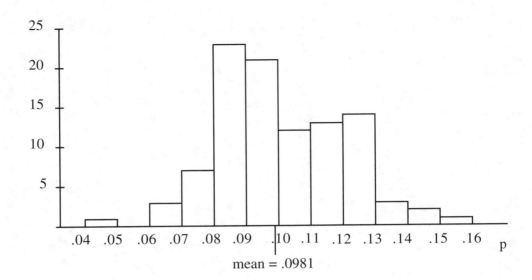

The statistic appears to have small bias as an estimate of the population proportion of 0.10.

(d) The mean of this distribution is the same as the mean for the population, namely 0.10, since the samples are simple random samples for which the sample proportion $\hat{p}$ is unbiased.

(e) Once again, since the samples are simple random samples (and hence the sample proportion is an unbiased estimate of the population proportion) the mean of the sampling distribution of $\hat{p}$ would be the same as for the population, namely 0.10. The spread would be smaller than that in our histogram of part (a). Larger samples give smaller spread.

Exercise 4.13

(a) Since all the probabilities must sum to 1 and the sum of the probabilities displayed is 0.9, the probability of drawing a tan candy must be 0.1.

(b) Since none of the events red, yellow, or orange have any outcomes in common, we can sum their probabilities to get the probability that the candy drawn is any of red, yellow, or orange. The value of this sum is 0.2 + 0.2 + 0.1 = 0.5.

Exercise 4.17

Assuming that the size of the sample is small compared to the sizes of the populations of New Jersey and the entire United States, the standard deviation would be about the same. Remember, the variability in the sample depends on the size of the sample, not on the size of the population from which the sample was drawn, provided the size of the population is much larger than the size of the sample.

Exercise 4.21

(a) The probability it is not forested is 1 minus the probability that it is forested, which is $1 - 0.35 = 0.65$.

(b) Since these two events have no outcomes in common (a forested acre is not pasture), the probability is the sum of the individual probabilities, namely $0.35 + 0.03 = 0.38$.

(c) The probability that the randomly chosen acre is something other than forest or pasture is 1 minus the probability that it is either forest or pasture. The probability that it is either forest or pasture was found in (b) to be 0.38. Thus the probability that the randomly chosen acre is something other than forest or pasture is $1 - 0.38 = 0.62$.

SECTION 4.2

(optional)

SECTION OVERVIEW

This section discusses details of probability. The statistics we are studying are also called **random variables** and their distribution is called a **probability distribution** because the long-run frequency of occurrences of each value is its probability. There are two types of random variables: **discrete** and **continuous**. The **law of large numbers** describes how the mean of many observations of a random process will get closer and closer to the mean of the variable. This section also shows that the normal distribution is a type of probability distribution.

KEY CONCEPTS

Details on discrete random variables

- A discrete random variable has a limited, finite, number of possible values.

- The distribution can be displayed as a probability histogram.

- The mean, μ is the sum of each possible value, x, multiplied by its probability, p.

  Here is the formula we will use:

$$\mu = x_1 p_1 + x_2 p_2 + \ldots + x_n p_n$$

- The variance is the average of the squared deviations of each value from the mean multiplied by the probability of each value. Or in formula notation:

$$\sigma^2 = (x_1 - \mu)^2 p_1 + (x_2 - \mu)^2 p_2 \ldots + (x_n - \mu)^2 p_n$$

- The standard deviation is the square root of the variance, just as it was in Chapter 1.

Details on continuous random variables

A continuous random variable has an infinite range (continuum) of values that the variable could take. The probability distribution assigns probabilities to a range of values rather than a single value. The distribution can be described as a density curve.

SOLUTIONS TO SELECTED TEXT EXERCISES

Exercise 4.25

(a) The percent of fifth-graders that eventually finished twelfth grade is given by the probability that X is 12. This probability can be read directly from the table as 0.752.

(b) We notice that all the probabilities given are between 0 and 1. If we add up the probabilities listed we get

$$0.010 + 0.007 + 0.007 + 0.013 + 0.032 + 0.068 + 0.070 + 0.041 + 0.752 = 1$$

Thus this is a legitimate discrete probability distribution. A probability histogram is given below.

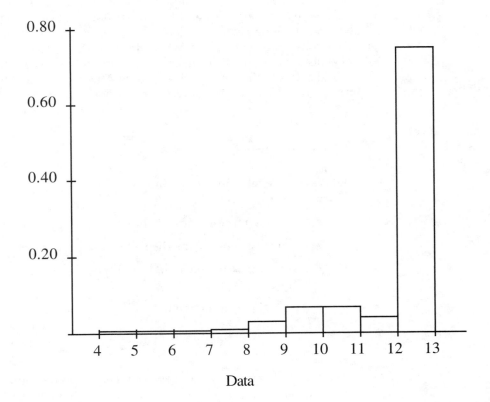

(c) $P(X \geq 6)$ is the sum of all the probabilities.

$$P(X = 6) + P(X = 7) + P(X = 8) + P(X = 9) + P(X = 10) + P(X = 11) + P(X = 12)$$
$$= 0.007 + 0.013 + 0.032 + 0.068 + 0.070 + 0.041 + 0.752 = 0.983$$
$$= 0.983$$

(d) $P(X > 6)$ is the sum of all the probabilities.

$$P(X = 7) + P(X = 8) + P(X = 9) + P(X = 10) + P(X = 11) + P(X = 12)$$
$$= 0.007 + 0.013 + 0.032 + 0.068 + 0.070 + 0.041 + 0.752 = 0.983$$
$$= 0.976$$

(e) The values of X that make up the event "The student completed at least one year of high school" are the values $X = 9$, $X = 10$, $X = 11$, and $X = 12$. Thus the probability of this event is

$$P(X = 9) + P(X = 10) + P(X = 11) = P(X = 12) \quad = 0.068 + 0.070 + 0.041 + 0.752$$
$$= 0.931$$

Exercise 4.29

(a) There are 10,000 four-digit numbers 0000 to 9999.

(b) If the plates are assigned at random, all four-digit numbers are equally likely. The probability of getting any specific four-digit number is

$$1/(\text{total number of four-digit numbers})$$

or $1/10,000 = 0.0001$.

(c) There are ten four-digit numbers with all four digits identical. These are 0000, 1111, 2222, 3333, 4444, 5555, 6666, 7777, 8888, and 9999. The probability that your plate has such a number is then

$$\frac{\text{count of all outcomes producing all four digits identical}}{\text{total number of four digit numbers}} = \frac{10}{10,000} = \frac{1}{1,000} = 0.001$$

Exercise 4.31

The ten values are equally likely and so each has probability 0.1. Thus the mean is

$$\mu = 0 \times 0.1 + 1 \times 0.1 + 2 \times 0.1 + 3 \times 0.1 + 4 \times 0.1 + 5 \times 0.1 + 6 \times 0.1 + 7 \times 0.1 + 8 \times 0.1 + 9 \times 0.1$$
$$= 0 + 0.1 + 0.2 + 0.3 + 0.4 + 0.5 + 0.6 + 0.7 + 0.8 + 0.9$$
$$= 4.5$$

This result could be seen from the histogram in Figure 4.7(b) since the histogram is symmetric. The mean must therefore be in the center of the histogram at 4.5.

Exercise 4.35

(a) The density curve is reproduced below.

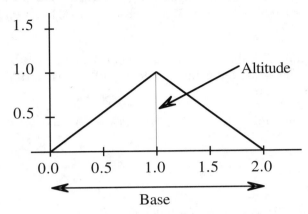

Recall that the area of a triangle is (1/2)(height of altitude)(length of base). In this triangle, the altitude has height 1 and the base has length 2. Thus the area is (1/2)(1)(2) = 1.

(b) The desired probability is the area of the shaded portion of the triangle below.

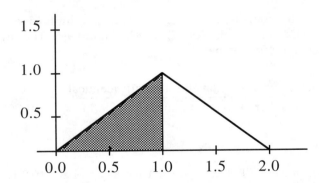

The shaded portion is itself a triangle with altitude having height 1 and base having length 1. Thus the area of the shaded portion is (1/2)(height of altitude)(length of base) = (1/2)(1)(1) = 1/2. The probability that Y is less than 1 is therefore 1/2.

(c) The desired probability is the area of the shaded portion of the triangle below.

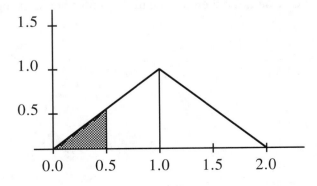

The shaded portion is itself a triangle with altitude having height 1/2 and base having length 1/2. Thus the area of the shaded portion is (1/2)(height of altitude)(length of base) = (1/2)(1/2)(1/2) = 1/8. The probability that Y is less than 0.5 is therefore 1/8.

Exercise 4.39

First we calculate the mean, which is obtained by multiplying each possible value by its probability and adding over all the values. This gives

$$\mu = 4(.010) + 5(.007) + 6(.007) + 7(.013) + 8(.032) + 9(.068) + 10(.070) + 11(.041) + 12(.752)$$
$$= .040 + .035 + .042 + .091 + .256 + .612 + .700 + .451 + 9.024$$
$$= 11.251$$

To calculate the standard deviation we set up a table.

x_i	p_i	$(x_i - \mu)^2 p_i$	
4	.010	$(4 - 11.251)^2(.010) =$	.5258
5	.007	$(5 - 11.251)^2(.007) =$	.2735
6	.007	$(6 - 11.251)^2(.007) =$	.1930
7	.013	$(7 - 11.251)^2(.013) =$	.2349
8	.032	$(8 - 11.251)^2(.032) =$	.3382
9	.068	$(9 - 11.251)^2(.068) =$	.3446
10	.070	$(10 - 11.251)^2(.070) =$	.1096
11	.041	$(11 - 11.251)^2(.041) =$	.0026
12	.752	$(12 - 11.251)^2(.752) =$	.4219
		$\sigma^2 =$	2.4441

Thus the standard deviation is $\sigma = \sqrt{2.4441} = 1.5634$.

Exercise 4.41

(a) $P(\hat{p} \geq .16) = P\left(\dfrac{\hat{p} - .15}{.0092} \geq \dfrac{.16 - .15}{.0092}\right) = P(Z \geq 1.09) = 1 - P(Z \leq 1.09) = 1 - .8621$
$$= .1379$$

A graph indicating the desired area under the standard normal curve is given below.

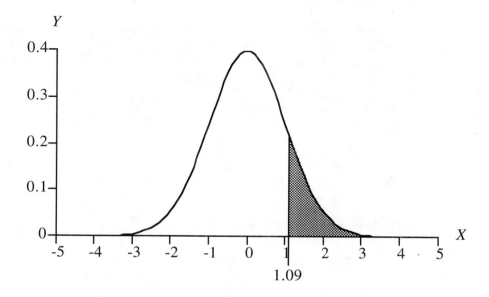

(b) $P(.14 \le \hat{p} \le .16) = P\left(\dfrac{.14-.15}{.0092} \le \dfrac{\hat{p}-.15}{.0092} \le \dfrac{.16-.15}{.0092}\right) = P(-1.09 \le Z \le 1.09)$

$= P(Z \le 1.09) - P(Z \le -1.09) = .8621 - .1379 = .7242.$

A graph indicating the desired area under the standard normal curve is given below.

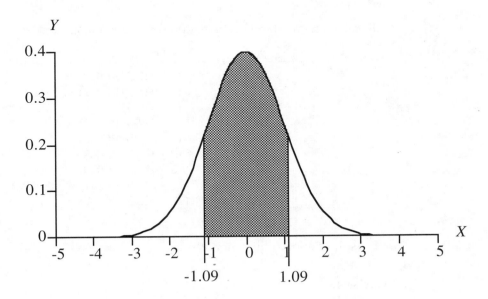

SECTION 4.3

SECTION OVERVIEW

Section 4.3 gives details about using a specific statistic, the **sample proportion,** $\hat{p}$. We learn about the sampling distribution of the statistic and the mean and standard deviation. In this section we also learn how to use the **normal distribution to approximate the sampling distribution** of $\hat{p}$. These calculations are the same as the ones we did in Chapter 1. Finally, we see how to use sample proportions to solve problems about the **sample count,** by restating the problem in terms of the sample proportion.

KEY CONCEPTS

Details on the sampling distribution of $\hat{p}$

- The shape of the distribution is close to normal and gets more and more normal as the sample size, n, increases. Use the normal approximation only if $np \ge 10$ and $n(1-p) \ge 10$.

- The mean of the distribution is the population proportion, p.

- The standard deviation gets smaller as the sample gets larger.

- The standard deviation can be calculated using the following formula:
$\sqrt{\dfrac{p(1-p)}{n}}$, where n is the sample size.
- Use this formula for the s.d. only when the population is at least 10 times larger than the sample.

Calculations with $\hat{p}$

Calculating the chance of $\hat{p}$ being in a certain range is done the same way as in Chapter 1. Now, however we can use probability notation.

1. *State the problem.* Using probability notation, we can write $P(a < \hat{p} < b)$ to mean "the probability that $\hat{p}$ is greater than a but less than b."

2. *Draw a picture of the problem.* It will help you to know what area you are looking for.

3. *Standardize the observations.* You may need to calculate the standard deviation using the formula presented before.

4. *Use the table.* Using Table A on page 626, find the area you need.

SOLUTIONS TO SELECTED TEXT EXERCISES

Exercise 4.47

(a) The sample proportion is $\hat{p}$ = 86/100 = .86.

(b) We want to determine the probability that the sample proportion $\hat{p}$ resulting from an SRS of size 100 from a population in which the true proportion is p = .90 would be .86 or smaller. In symbols, we want $P(\hat{p} \leq .86)$. Since $np = 100(.90) = 90$ and $n(1 - p) = 100(1 - .90) = 10$ are both ≥ 10, and since the population size of 5000 is at least 10 times the sample size n = 100, we may use the normal approximation to the sampling distribution of $\hat{p}$. The normal approximation says that the sampling distribution of $\hat{p}$ is approximately normal with mean $\mu = p = .90$ and standard deviation $\sigma = \sqrt{p(1-p)/n} = \sqrt{(.90)(1-.90)/100} = .03$. Thus

$$P(\hat{p} \leq .86) = P\left(\dfrac{\hat{p}-.9}{.03} \leq \dfrac{.86-.90}{.03}\right) = P(Z \leq -1.33) = .0918$$

A graph indicating the desired area under the standard normal curve is given below.

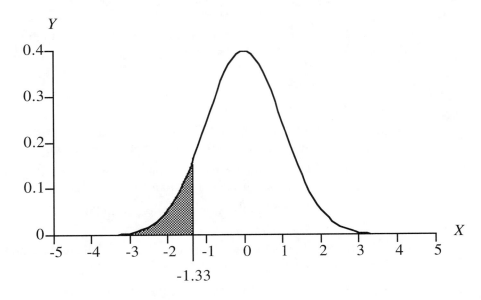

-1.33

(c) The on-time percentage in our sample was 86%. While this is smaller than 90%, we must remember that samples from a population are subject to sampling variability. They do not always give a value of a sample statistic that is exactly equal to the corresponding population parameter. In repeated SRS's of size 100 from a population in which the true proportion is 90%, sometimes we will get a sample proportion less than 90% and sometimes more than 90%. How likely the sample proportion is to differ from the population proportion by a given amount is determined by the sampling distribution of the sample proportion. We must ask whether the sample proportion of 86% is very unlikely to occur in an SRS of size 100 from a population of 5000 in which the population proportion is 90%. Part (b) shows that the probability of obtaining a sample proportion as small as 86% or smaller when taking an SRS of size 100 from a population of 5000 in which the population proportion is 90% is .0918. While small, this is not so small as to lead us to believe that the sample proportion of 86% is very unlikely to have arisen by chance. Thus our sample result is not sufficiently small to convincingly refute the 90% claim.

Exercise 4.49

(a) The mean number who will accept admission is $\mu = np = (1700)(.70) = 1190$.

(b) Here we wish to calculate the probability that fewer than 1200 students accept, or less than or equal to 1199 accept. We need to restate this count in terms of a proportion. Since the sample size is 1700, the count of 1199 is equivalent to the proportion 1199/1700 = .705. Thus the probability that fewer than 1200 students accept is the same as the probability that $\hat{p}$ = the proportion of students that accept, is less than or equal to .705. We note that the supposed population size of all qualified students is likely to be several hundred thousand, which is at least ten times the sample size of 1700. In addition we check that $np = (1700)(.70) = 1190$ and $n(1 - p) = (1700)(1 - .70) = (1700)(.30) = 510$ are both at least ten. Thus we may safely use the normal approximation to compute probabilities involving the sample proportion $\hat{p}$. The sampling distribution of $\hat{p}$ is approximately normal with mean .70 and standard deviation $\sqrt{p(1-p)/n}$ $= \sqrt{.7(1-.7)/1700}$ $= .011$.

The desired probability is thus

$$P(\hat{p} \le .705) = P\left(\frac{\hat{p}-.7}{.011} \le \frac{.705-.7}{.011}\right) = P(Z \le 0.45) = .6736.$$

A graph illustrating the desired area under the standard normal curve is given below.

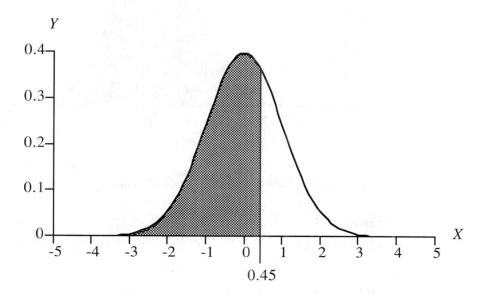

0.45

The probability that fewer than 1200 students accept is .6736.
Note that you might ignore the distinction of "fewer than 1200" and determine the probability that $\hat{p}$ is less than 1200/1700=.706. Your answer will be slightly larger than the .6736 we found above.

(c) We must restate the counts in terms of proportions. 1150 is 1150/1700 = .676 and 1250 is 1250/1700 = .735 of the sample of 1700 students. In terms of proportions, we want to know the probability that the sample proportion $\hat{p}$ is between .676 and

.735. As we saw in part (b), the sampling distribution of $\hat{p}$ is approximately normal with mean .70 and standard deviation .011. The desired probability is thus

$$P(.676 \le \hat{p} \le .735) = P\left(\frac{.676-.7}{.011} \le \frac{\hat{p}-.7}{.011} \le \frac{.735-.7}{.011}\right) = P(-2.18 \le Z \le 3.18)$$

$$= P(Z \le 3.18) - P(Z \le -2.18) = .9993 - .0146 = .9847.$$

A graph illustrating the desired area under the standard normal curve is given below.

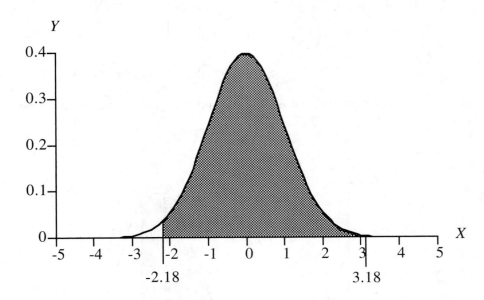

The probability that between 1150 and 1250 accept is .9847.

Exercise 4.53

(a) The mean number of heart attacks in one group of 2000 men is

$$\mu = np = (2000)(.04) = 80$$

(b) We wish to calculate the probability that the group will suffer at least 75 heart attacks. We need to restate this count in terms of a proportion. Since the sample size is 2000, the count of 75 is equivalent to the proportion 75/2000 = .0375. Thus the probability that the group will suffer at least 75 heart attacks is the same as the probability that $\hat{p}$ = the proportion of the group that suffer a heart attack, is greater than or equal to .0375. We note that the supposed population size is large (the population is apparently adult men over the age of about 40) and at least ten times the sample size of 2000. In addition we check that $np = (2000)(.04) = 80$ and $n(1 - p) = (2000)(1 - .04) = (2000)(.96) = 1920$ are both at least ten. Thus we may safely use the normal approximation to compute probabilities involving the sample proportion $\hat{p}$. The sampling distribution of $\hat{p}$ is approximately normal with mean .04 and standard deviation $\sqrt{p(1-p)/n} = \sqrt{.04\,(1-.04\,/\,2000} = .0044$.

The desired probability is thus

$$P(\hat{p} \geq .0375) = P\left(\frac{\hat{p}-.04}{.0044} \geq \frac{.0375-.04}{.0044}\right) = P(Z \geq -0.57) = .7157$$

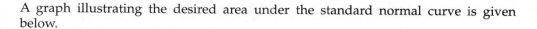

A graph illustrating the desired area under the standard normal curve is given below.

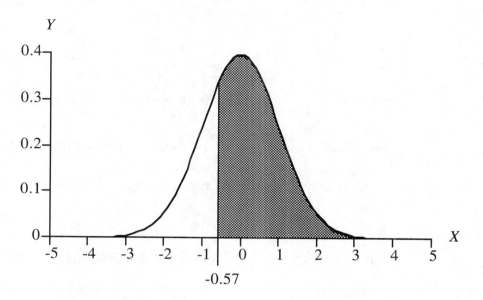

The probability that the group will suffer at least 75 heart attacks is .7157.

SECTION 4.4

(optional)

SECTION OVERVIEW

In Section 4.3 we saw that the sampling distribution of the sample proportion and the sample count are approximately normal under certain conditions. In this optional section we will see what the distribution is exactly, when the setting meets certain conditions called the **binomial setting**. The distribution is called the **binomial distribution**. The binomial distribution can be used to calculate probabilities.

KEY CONCEPTS

The binomial setting

In order to describe the distribution of the sample count as binomial the situation must fit the following set up:

- There are n observations.

- The observations are independent.

- Each observation is classified as one of two categories, usually called a "success" or a "failure."

- The probability of a success, called p, is the same for each observation.

Details on the binomial distribution

If X is the number of successes in n trials (observations), then the probability of having k success can be calculated as follows:

$$P(X = k) = \binom{n}{k} p^k (1-p)^{n-k}$$

where $\binom{n}{k} = \dfrac{n!}{k!(n-k)!}$ and $n! = n \times (n-1) \times (n-2)\ldots 3 \times 2 \times 1$.

$n!$ is called "n factorial" and $\binom{n}{k}$ is called the binomial coefficient. k can be any number from 0 to n and $0! = 1$.

The mean of a binomial random variable is $\mu = np$ and the standard deviation is $\sigma = \sqrt{np(1-p)}$.

SOLUTIONS TO SELECTED TEXT EXERCISES

Exercise 4.57

X, the number of problems the student gets right, probably does not have a binomial distribution. Since additional instruction is given between problems that the student gets wrong, it is likely that this instruction increases the probability that the student gets the next question correct. The probability p of a success (getting a question right) is probably not the same for each observation (question) and so X is probably not binomial.

Exercise 4.59

(a) X counts the number of children, out of 5, who have type O blood. The possible values of X are therefore the integers 0, 1, 2, 3, 4, and 5.

(b) Using the formula for binomial probabilities we calculate

$$P(X = 0) = \binom{5}{0}(.25)^0(.75)^5 = \frac{5!}{0!5!}(1)(.2373) = (1)(1)(.2373) = .237$$

$$P(X = 1) = \binom{5}{1}(.25)^1(.75)^4 = \frac{5!}{1!4!}(.25)(.3164) = (5)(.25)(.3164) = .396$$

$$P(X = 2) = \binom{5}{2}(.25)^2(.75)^3 = \frac{5!}{2!3!}(.0625)(.4219) = (10)(.0625)(.4219) = .264$$

$$P(X = 3) = \binom{5}{3}(.25)^3(.75)^2 = \frac{5!}{3!2!}(.0156)(.5625) = (10)(.0156)(.5625) = .088$$

$$P(X = 4) = \binom{5}{4}(.25)^4(.75)^1 = \frac{5!}{4!1!}(.0039)(.75) = (5)(.0039)(.75) = .015$$

$$P(X = 5) = \binom{5}{5}(.25)^5(.75)^0 = \frac{5!}{5!0!}(.0010)(1) = (1)(.0010)(1) = .001$$

A probability histogram representing this distribution is given below.

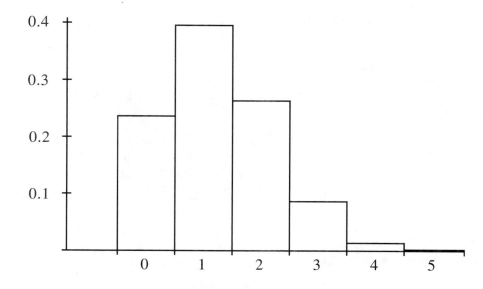

Exercise 4.63

The mean μ and standard deviation σ are

$$\mu = np = (5)(.25) = 1.25$$

$$\sigma = \sqrt{np(1-p)} = \sqrt{5(.25)(.75)} = .9682$$

We mark the mean on the probability histogram reproduced from Exercise 4.59.

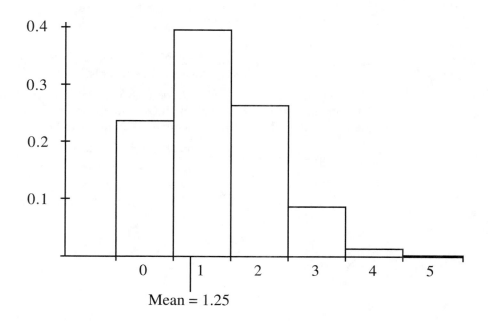

Mean = 1.25

Exercise 4.67

(a) The total number of guesses is 20, so $n = 20$. The probability of a correct guess is 0.25, so $p = .25$.

(b) The mean number of correct guesses is $\mu = np = (20)(.25) = 5$.

(c) The probability of exactly 5 correct guesses is, using the binomial probability formula,

$$\binom{20}{5}(.25)^5(.75)^{15} = \frac{20!}{5!15!}(.0009766)(.0133635)$$

$$= \frac{(20)(19)(18)(17)(16)}{(5)(4)(3)(2)(1)}(.0009766)(.0133635)$$

$$= (15504)(.0009766)(.0133635)$$

$$= .20234$$

Exercise 4.71

(a) Since all outcomes are equally likely,

$$P(\text{card drawn is black}) = \frac{\text{number of black cards}}{\text{total number of cards}} = \frac{26}{52} = .5$$

$$P(\text{card drawn is red}) = \frac{\text{number of red cards}}{\text{total number of cards}} = \frac{26}{52} = .5$$

(b) If the first card drawn is black, 51 cards remain in the deck and 25 of them are black. The remaining 26 cards are red. If we now draw a second card without returning the first to the deck, all 51 cards are equally likely to be drawn. Thus

$$P(\text{card drawn is black}) = \frac{\text{number of black cards}}{\text{total number of cards}} = \frac{25}{51} = .4902$$

$$P(\text{card drawn is red}) = \frac{\text{number of red cards}}{\text{total number of cards}} = \frac{26}{51} = .5098$$

(c) If the first card drawn was red, 51 cards remain in the deck and 25 of them are red. The remaining 26 cards are black. If we now draw a second card without returning the first to the deck, all 51 cards are equally likely to be drawn. Thus

$$P(\text{card drawn is black}) = \frac{\text{number of black cards}}{\text{total number of cards}} = \frac{26}{51} = .5098$$

$$P(\text{card drawn is red}) = \frac{\text{number of red cards}}{\text{total number of cards}} = \frac{25}{51} = .4902$$

SECTION 4.5

SECTION OVERVIEW

Earlier in this chapter we looked at the sampling distribution of the proportion of individuals in a sample that fall into a certain category. In this section we will look at the sampling distribution of the mean of the responses from an SRS. Two principles about means are important: means have less variability than the individual observations the means are calculated from, and the means are more normal than the individual observation. The **sample mean,** $\overline{x}$, is an estimate of the unknown **population mean,** μ. As with all sampling distributions we need to know the mean and standard deviation, and the shape of the distribution. In fact, we know from the **central limit theorem** that the shape of the distribution of the sample mean is very close to normal under certain circumstances. The **law of large numbers** also tells us more about the behavior of $\overline{x}$ as the sample size increases.

KEY CONCEPTS

Details on the sample mean $\overline{x}$

- $\overline{x}$ is an unbiased estimate of the population mean μ, so the mean of $\overline{x}$ is μ, the population mean.

- The standard deviation is $\sigma/\sqrt{n}$, where σ is the population standard deviation. Only use this formula if the population is at least 10 times larger than the sample.

- This is true no matter what the shape of the population distribution.

- The shape of the distribution of the sample mean depends on the shape of the population distribution. If the population was normal, $N(\mu,\sigma/\sqrt{n})$, then the sample mean has normal distribution, $N(m,s/\sqrt{n})$.

More about the central limit theorem

The central limit theorem tells us about the shape of the sampling distribution of the mean when the sample is an SRS and n is large. The population can have any mean, μ, but the standard deviation, σ, must be finite. When this condition is met, the theorem says if n is large then the distribution of the sample mean will be close to normal, $N(\mu,\sigma/\sqrt{n})$, regardless of the shape of the population distribution.

This theorem lets us perform probability calculations about the sample mean just like the ones we did in Chapter 1 and with the sample probability earlier in the chapter.

More about the law of large numbers

The law of large numbers describes how the sample mean, $\overline{x}$, of many observations of a population will get closer and closer to the true mean, μ, as the sample size gets larger.

SOLUTIONS TO SELECTED TEXT EXERCISES

Exercise 4.73

The mean of the sampling distribution of the mean returns $\overline{x}$ for all possible samples of 5 stocks is the same as for the population of all stocks, namely $\mu = -3.5\%$. The standard deviation of the sampling distribution of $\overline{x}$ is $\sigma/\sqrt{n}$, where σ is the standard deviation of the population, 26%. Thus

standard deviation of the sampling distribution of $\overline{x}$ = $\sigma/\sqrt{n}$ = $\dfrac{26\%}{\sqrt{5}}$ = 11.6276%

Exercise 4.79

(a) Let $\overline{x}$ denote the mean of 3 weighings. The sampling distribution of $\overline{x}$ is normal (since the distribution of individual weighings is normal) with

mean = the mean of the distribution of individual weighings = 123

$$\text{standard deviation} = \frac{\text{standard deviation of population of individual weighings}}{\sqrt{\text{sample size}}}$$

$$= .08/\sqrt{3} = .0462$$

(b) The probability that a particular average of 3 weighings $\overline{x}$ will have a weight of 124 mg or higher is (since $\overline{x}$ is normal with mean 123 and standard deviation .0462)

$$P(\overline{x} \geq 124) = P(\frac{\overline{x} - 123}{.0462} \geq \frac{124 - 123}{.0462}) = P(Z \geq 21.65)$$

$$= 0 \text{ (actually a value smaller than .0003. 21.65 is off our table)}$$

Exercise 4.85

(a) The distribution of a single measurement on Judy is normal with mean $\mu = 3.8$ and standard deviation $\sigma = 0.2$. Judy is diagnosed as hypokalemic if a single measurement X is below 3.5. The probability of obtaining such a measurement is

$$P(X \leq 3.5) = P\left(\frac{X - 3.8}{0.2} \leq \frac{3.5 - 3.8}{0.2}\right) = P(Z \leq -1.50) = .0668$$

A graph illustrating the desired area under the standard normal curve is given below.

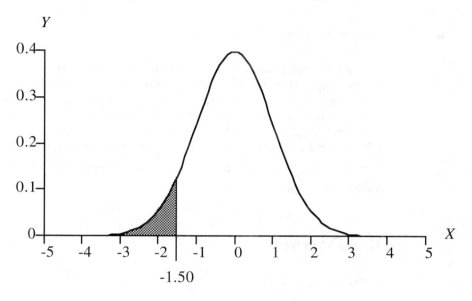

-1.50

(b) We assume the fact that the measurements are made on separate days makes it reasonable to assume the measurements can be regarded as an SRS of size 4. Then the distribution of the mean of 4 such measurements is normal with the same mean as for a single measurement, $\mu = 3.8$, and standard deviation $\sigma/\sqrt{4} = 0.2/2 = 0.1$. Judy is diagnosed as hypokalemic if a the mean of 4 measurements $\overline{X}$ is below 3.5.

The probability of obtaining such a measurement is

$$P(\overline{X} \le 3.5) = P\left(\frac{\overline{X} - 3.8}{0.1} \le \frac{3.5 - 3.8}{0.1}\right) = P(Z \le -3.00) = .0013$$

A graph illustrating the desired area under the standard normal curve is given below.

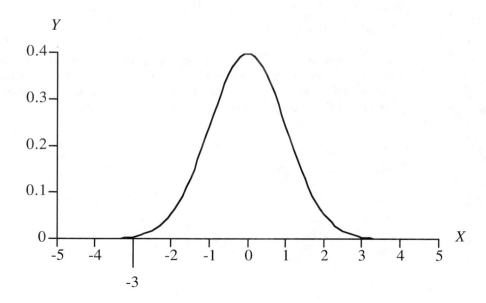

-3

SECTION 4.6

(optional)

Section Overview

As we have seen, the only data that have a variance of zero are those whose values are the same. This is extremely rare in real life. So how do we tell the natural variation in measurements from the big deviations that signal something has changed? One way is to use **control charts**. These charts are most commonly used to monitor manufacturing processes. Since variation is always present, we create these charts with **control limits** to signal us if the process changes too much. The two most popular charts are the $\bar{x}$ **chart** and the $\hat{p}$ **chart**. These charts are really two specialized forms of time plots.

Key Concepts

Details on the $\bar{x}$ chart

Using what we learned about $\bar{x}$ earlier in the chapter and applying the 68–95–99.7% rule from Chapter 1, we can create a chart that shows when a process is out of control.

To create an $\bar{x}$ chart we need to know the process mean and the standard deviation of the sample mean. The population mean, m, is usually unknown. If this is the case, a historical mean may be used. This mean is the center line of the chart. At the top and bottom of the graph is a control limit. These limits are usually set at m $\pm$ s/$\sqrt{n}$, which will capture 99.7% of the sample means.

Plot the sample means, $\bar{x}$, from samples of size n, against time. The center line is μ, and the limits are set as above.

"Out of control" is declared if a sample mean goes outside the control limits or if a pattern appears in the chart.

Details on the $\hat{p}$ chart

The $\hat{p}$ chart is basically the same as the $\bar{x}$ chart except we plot the sample proportion instead of the sample mean. Again, we need to know the sampling distribution of the statistic we are plotting to create the control limits and the center line. The center line of the chart is p and the control limits are set at $p \pm 3\sqrt{p(1-p)/n}$.

SOLUTIONS TO SELECTED TEXT EXERCISES

Exercise 4.89

Since the target value of $\mu = .8750$ inch is the mean of slots produced by the milling machine when properly adjusted, the center line is the horizontal line at $\mu = .8750$. The standard deviation of slots produced by the milling machine when properly adjusted is $\sigma = .0012$. For means based on samples of size 5, the control limits are horizontal lines at

$$\mu \pm 3\sigma / \sqrt{n} = .8750 \pm 3(.0012)/ \sqrt{5} = .8750 \pm .0016$$

or at .8734 and .8766.

Exercise 4.91

When the process is properly adjusted, the mean is $\mu = 2.2050$ cm with standard deviation $\sigma = .0010$ cm. For means based on samples of size 5, the center line is the horizontal line at $m = 2.2050$ and the control limits are horizontal lines at

$$\mu \pm 3\sigma / \sqrt{n} = 2.2050 \pm 3(.0010)/ \sqrt{5} = 2.2050 \pm .0013$$

or at 2.2037 and 2.2063. The $\bar{x}$ control chart is thus

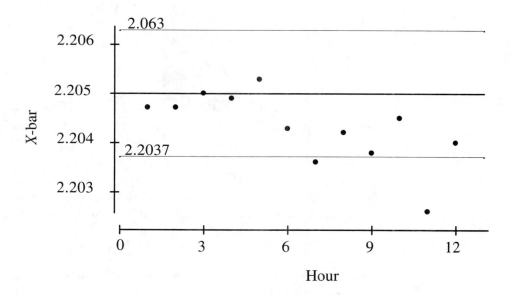

Using the "one point out" signal, we are alerted at samples 7 and 11 that the process appears out of control. Action should have been taken at sample 7. Using the "run of nine" signal, the process never appears out of control (although the last 8 points are all on the same side of the center line).

Exercise 4.95

(a) We want to know the probability that the line width X, which varies according to a normal distribution with mean $\mu = 2.829$ mm and standard deviation $\sigma = 0.1516$ mm, takes on a value below $3.0 - .2$ mm $= 2.8$ mm or above $3.0 + .2$ mm $= 3.2$ mm. This probability is

$$P(X \le 2.8) + P(X \ge 3.2) = P\left(\frac{X - 2.829}{0.1516} \le \frac{2.8 - 2.829}{0.1516}\right) + P\left(\frac{X - 2.829}{0.1516} \ge \frac{3.2 - 2.829}{0.1516}\right)$$

$$= P(Z \le -0.19) + P(Z \ge 2.45)$$

$$= .4247 + .0071$$

$$= .4318$$

This probability is displayed as the shaded portion of the normal curve below.

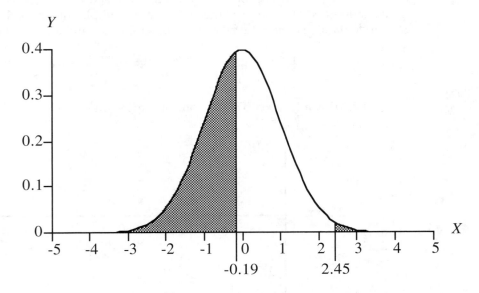

The percent of chips outside the acceptable range is therefore 43.18%.

(b) Assuming the target value of 3.0 is the value of the center line, μ, the control limits for an $\bar{x}$ chart for line width if samples of size 5 are used are

$$\mu \pm 3\sigma/\sqrt{n} = 3.0 \pm 3(.1516)/\sqrt{5} = 3.0 \pm .2034$$

Thus the lower control limit is the horizontal line at 2.7966 and the upper control limit is the horizontal line at 3.2034.

SOLUTION TO SELECTED TEXT REVIEW EXERCISES

Exercise 4.103

(a) Since scores are approximately normally distributed with mean $\mu = 100$ and standard deviation $\sigma = 15$, the probability that a randomly chosen individual has a score X of 105 or higher is

$$P(X \geq 105) = P\left(\frac{X - 100}{15} \geq \frac{105 - 100}{15}\right) = P(Z \geq 0.33) = 1 - P(Z \leq 0.33) = 1 - .6293$$

$$= .3707$$

This probability is indicated by the shaded region in the graph of the normal curve shown below.

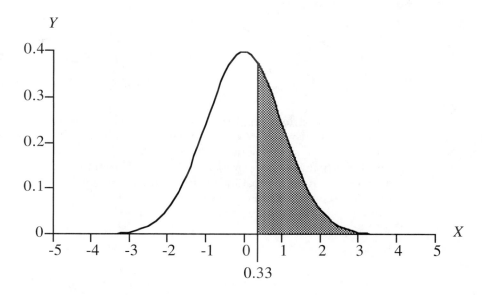

(b) For an SRS of 60 people, the mean and standard deviation of the average WAIS score $\overline{x}$ are

mean = mean of the distribution of individual scores = $\mu = 100$

standard deviation = $\sigma / \sqrt{n} = 15 / \sqrt{60} = 1.94$

(c) From (b) we know that the distribution of $\overline{x}$, the average WAIS score of an SRS of 60 people, is approximately normal (since the population of WAIS scores is approximately normal) with mean 100 and standard deviation 1.94. Thus the probability that $\overline{x}$ is 105 or higher is

$$P(\overline{x} \geq 105) = P\left(\frac{\overline{x} - 100}{1.94} \geq \frac{105 - 100}{1.94}\right) = P(Z \geq 2.58) = .0049$$

This probability is indicated by the shaded region in the graph of the normal curve displayed below.

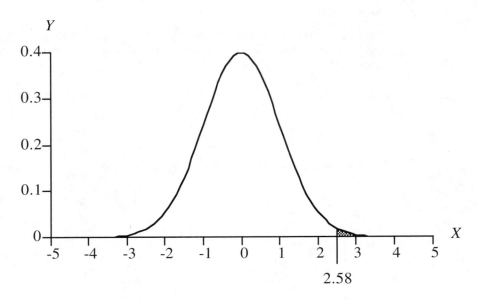

2.58

(d) If the distribution of WAIS scores in the adult population were distinctly non-normal, the answer to (a) would no longer be correct. The answer to (b) would remain correct, since the mean and standard deviation of the average do not depend on the normality of the population. We know from the central limit theorem, that the answer to (c) would most likely still be approximately correct. Unless the population of WAIS scores is extremely nonnormal, the distribution of the average of 60 scores from an SRS should be approximately normal.

Exercise 4.107

We know that the sampling distribution of $\hat{p}$ is approximately normal with mean μ $= p$ and standard deviation $\sigma = \sqrt{p(1-p)/n}$. If the coin is perfectly balanced, $p = 0.5$

so that $\mu = 0.5$ and $\sigma = \sqrt{(0.5)(1 - 0.5)/10,000} = \sqrt{.000025} = .005$. The probability that a balanced coin would give 5,067 or more heads in 10,000 tosses is the same as the probability that $\hat{p} \geq 5,067/10,000 = .5067$ and therefore approximately

$$P(\hat{p} \geq .5067) = P\left(\frac{\hat{p} - .5}{.005} \geq \frac{.5067 - .5}{.005}\right) = P(Z \geq 1.34) = .0901$$

This probability is indicated by the shaded region in the graph of the normal curve displayed below.

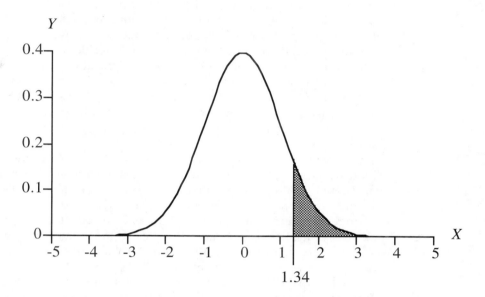

This probability .0901, means that there is only a 9.01% chance that one would obtain 5,067 or more heads in 10,000 tosses of a perfectly balanced coin. While this probability is fairly small, it is not so extreme as to be unreasonable if the coin was perfectly balanced. We would conclude that the evidence is not strong enough to doubt that the coin was perfectly balanced. Note, however, that the evidence is also consistent with the assumption that the coin is slightly unbalanced in favor of heads.

CASE STUDY

Our intuition about probability is often quite poor. Think of all the people that gamble and how few actually win. Psychologists confirm that probability is a difficult concept. Even after mastering the material in this chapter, you may still agree. To illustrate the difficulties even experts experience, we look at a version of a puzzle that appeared in Parade magazine in the column "Ask Marilyn." This puzzle was posed by Marilyn Savant, who claims to have the world's highest IQ. It is a variation on an old problem called the "Prisoner's Dilemma." This problem and it's solution generated months of controversy, with even experts disagreeing on the correct answer!

The setting is a quiz show, like the old show "Lets Make a Deal." You are the contestant. The host presents you with three doors. Behind one of them is a sports car. Behind the other two, goats. We assume that you want the sports car and are not interested in the goats. The host knows which door the sports car is behind, but you don't. You are invited to pick a door. After you pick, the host opens one of the two doors you did not select. Behind it is a goat. The host then asks you if you would like to switch the door you picked with the door the host did not open. Should you switch? An important fact to keep in mind is that the host, who knows which door hides the sports car, *always opens a door with a goat behind it* before offering you the opportunity to switch (after all, if the host opened the door with the sports car, the game would be over. There would be no point in switching!). Some of the controversy over this problem, which became quite heated, was due to a misunderstanding of this point. Before you continue reading, decide whether you would switch.

Here are some lines of reasoning you might adopt in deciding if you should switch. One possibility is that you decide not to switch because you feel that you are being tempted by the host. If you switch, you just know that it will turn out to be a mistake. This is based on intuition rather than calculation, and is not a good basis for a decision (again, think of all the people that gamble. How many play hunches? How many really win?).

Another line of reasoning might be the following. After the host opens a door, behind which is a goat, only two doors remain. One has the sports car, the other a goat. There is a 50 - 50 chance that the door you picked has the sports car. Therefore it shouldn't matter whether or not you switch. The odds are the same for switching as for not switching. This reasoning sounds plausible. It also sounds like the kind of trick answer faculty usually give students. Unfortunately it is wrong, but it is not easy to see why. Correct calculation of the probabilities is complicated. We will indicate below how you can estimate the correct probability on your own. First, however, we will reveal the correct answer and try to make the answer plausible.

It turns out that there is a 1/3 probability that the door you selected has the sports car and a 2/3 probability that the door you can switch to has the car. Thus, the best strategy is to switch. To see why, consider the following. At the outset, it seems reasonable to assume that the car is equally likely to be behind any of the doors. At least that is the premise of the show. This means that there is a 1/3 chance that the door you select has the car. So far, so good. Notice this also means that there is a 2/3 chance that the car is behind one of the other two doors. If at this point the host gave you the opportunity to trade the door you picked for *both* of the other doors (a 1 for 2 swap), it would make sense to switch. You would have a 2/3 chance of having the door with the car rather than a 1/3 chance. If you think about it, this is actually what you are being offered by the host when you are given the opportunity to switch. The host always opens a door with a goat and so is showing you which of the two doors you don't want. In essence, when you switch you are getting both doors (at least one of which must have a goat behind it) and are then having one with a goat behind it removed from further consideration. Switching is like getting two doors and hence has a 2/3 chance of yielding you the sports car.

The above explanation may leave you unconvinced. You may, as many did when this puzzle was first posed, still feel that the first explanation (there is a 50 - 50 chance the door you first picked has the car) is plausible. It turns out that part of the confusion arises from ignoring the fact that the host *always opens a door with a goat.* This must be taken into account in any correct calculation of the probabilities.

We need not carry out complicated calculations to determine whether there is a 50 - 50 chance that the door you first chose had the sports car or only a 1/3 chance. Instead, we can carry out a simulation of the quiz show and use the notion that probability is the proportion of times an outcome occurs in a large number of trials, to estimate the desired probability. By simulating the quiz show many times and keeping track of the proportion of times the door you first chose had the sports car, you can estimate this probability. Here is one way you might do the simulation. Have a friend roll a die. A 1 or 2 means the car is behind door #1. A 3 or 4 means the car is behind door #2 and a 5 or 6 means the car is behind door #3. Your friend should cover the die so you can't see the result (but he/she can). Pick a door. Your friend will then act as host and will reveal that a goat is behind one of the two remaining doors and offer you the opportunity to switch. For example, suppose your friend rolls a 3. This means the car is behind door #2. Suppose you pick door #1. There is a goat behind door #1 and door #3. Your friend will "open" door #3 (remember, of the doors you did not select, the host must always open one with a goat behind it) and tell you that there is a goat behind it, offering you the opportunity to switch to door #2. For purposes of the simulation, you should adopt the "refuse to switch" strategy every time. Repeat this simulation many times, keeping track of the results of the "refusal to switch strategy." The proportion of times that you win by refusing to switch would be an estimate of the probability that you will win by refusing to switch. One minus this proportion is an estimate of the probability that you would win by switching. Try this and see what you get.

Another way to do the simulation is to use the random digits table instead of a die. Begin by picking a random starting point in the table and read the table in single digits. Cover over the line you are to read and pick a door. You will adopt the strategy of refusing to switch. Now uncover the next digit in the random digits table. A 1, 2, or 3 indicates that the sports car is behind door #1, a 4, 5, or 6 door #2, and a 7, 8, or 9 door #3. Skip over 0s. You win if the digit indicates that the door you chose has the car. You lose if it indicates that the car is behind another door. Notice that if the car is behind a door other than the one you initially chose, the host would show you a goat behind a door you didn't choose and then offer you the opportunity to switch doors(which you would refuse). Repeat this simulation many times, keeping track of how often you win by refusing to switch doors. This proportion is an estimate of the probability that you win by refusing to switch. One minus this proportion is the probability that you would win if you switched doors. Does your simulation give an estimate of the probability that you would win by refusing to switch near 1/2 or 1/3?

Although using a simulation seems like a lazy way of avoiding doing the necessary calculations in order to determine the correct answer, and a method no self respecting scientist would adopt, simulations are very popular ways of calculating probabilities in complicated situations. Using a computer to run the simulation allows you to simulate a situation many, many times in a few seconds and get a good estimate of the desired probability as the long-run proportion of times the outcome of interest occurred in the simulation. Scientists and engineers routinely use simulations. You should not feel embarrassed by resorting to a simulation to answer the quiz show question above—it puts you in distinguished company.

CHAPTER 5

INTRODUCTION TO INFERENCE

CHAPTER OVERVIEW

After finishing four chapters of statistics and probability we are finally ready to make conclusions from the data and declare how sure we are about them. This is called formal inference. Everything we have covered so far leads us to inference. The methods we will learn are only valid if the data were collected from a random sample or a statistically planned randomized experiment, both of which were covered in Chapter 3. All the methods are based on the ideas of probability and sampling distributions from Chapter 4. Chapters 1 and 2 are the place to begin any exploration of data to see what methods we can use and to check the data for problems like outliers.

Inference means to conclude by reasoning. All the tools we will learn are based on the idea of what would happen if we used the tool many times under the same conditions. Inference often requires fancy looking formulae and many calculations. We aren't going to worry about those in this chapter. A computer or calculator can carry out the calculations. Chapter 5 will teach us the reasoning behind any inference technique. The ideas presented here are important because they are the foundation for all inference. We will use the sample mean as an example in this chapter but we wish to emphasize the ideas, not just the technical tools.

SECTION 5.1

SECTION OVERVIEW

Section 5.1 introduces the **confidence interval**. Confidence intervals are what we use when we estimate the value of an unknown population parameter. We use confidence intervals to say how sure we are that the unknown parameter lies between two values. All **intervals** have the form of **estimate ± margin of error**. The estimate is a statistic that we calculate from the sample and the margin of error is the "give or take" value, showing how accurate our estimate is. The margin of error is based on the standard deviation of the statistic and its distribution, and takes the form

$$z^* \sigma_{estimate}$$

where $\sigma_{estimate}$ is the standard deviation of the statistic. Along with the interval itself we give a probability. Using the probability rules that we have learned in past chapters, we state how sure we are about our interval; we state the **confidence level**, C. So, confidence intervals give us a range of values that may contain the unknown parameter and state how often, in the long run, the population parameter will be in the interval. The key to understanding a confidence interval is:

- A level C confidence interval says that the method we used will produce an interval which contains the true parameter with probability C.

- If we repeat this method again with a different sample we will produce an interval that contains the parameter C* 100% of the time.

Throughout this section we will demonstrate basic ways of calculating a confidence interval using the sample mean.

KEY CONCEPTS

The confidence level and critical values

Given the confidence level, C, we find the critical value, z^*, by looking up the z-score that has $(1 - C)/2$ in the tail area. The most common critical values are in the bottom line of table C. The critical value is identified by the probability in the tail. A 90% confidence interval has a 0.05 critical value. A 60% confidence interval has a 0.2 critical value.

$$C \rightarrow (1 - C)/2 \text{ area in the tail} = z^* \text{ value in the table}$$

The width of the interval is determined by the margin of error. If we are willing to accept a less confidence (i.e., a lower confidence level) then the critical value will be smaller and the interval smaller. The interval will also get smaller if the statistic has a smaller standard deviation.

Confidence intervals for the population mean

As an example we use the sample mean. The basic form of a C level confidence interval is estimate ± margin of error. For data from an SRS, the confidence interval about the population mean is then

$$\bar{x} \pm \sigma / \sqrt{n}$$

because $\bar{x}$ is the statistic we use to estimate the sample mean and $\sigma / \sqrt{n}$ is the standard deviation of the estimate.

Using confidence intervals

We will use a confidence interval (C.I.) to estimate the value of an unknown parameter. However, when planning a survey or experiment it pays to consider if you will be making any C.I.s. Knowing what sort of inference you want to do can help you choose the best sample size. Sample size, n, is in the formula for the standard deviation of almost any statistic we will see. Just look at the sample mean. If we want the margin of error to be a certain size, say m, then the sample size needs to be

$$n = \left(\frac{z \star 6}{m} \right)^2$$

Always round up if n is not a whole number. If n is unrealistic for your needs you may need to make the acceptable margin of error larger.

WARNING LABELS FOR C.I.s

- *Use the formulae presented in these chapters only if the data are from an SRS. Other sampling plans have different inference methods.*

- *Data must be collected using valid methods like those discussed in Chapter 3. Poor collection methods will give worthless results. Bias and variation caused by something other than chance is not accounted for by the margin of error.*

- *Apply the methods learned in Chapters 1 and 2 to be sure the data have no outliers, also check the population for skewness and non-normality. Use the tools we have already learned to decide if the inference will be valid.*

- *If the population is not normal and the sample size is small then these methods are not safe to use.*

SOLUTIONS TO SELECTED TEXT EXERCISES

Exercise 5.1

(a) Since the margin of error for a 95% confidence interval is ± 3%, the 95% confidence interval for the percent of all adult women who think they do not get enough time for themselves is 47% ± 3%, or between 44% and 50%.

(b) The value 47% is based on a sample of 1025 women selected at random from all women in the United States (excluding Alaska and Hawaii). Another sample of 1025 women might yield a different percent. Repeated random samples of 1025 women will yield a variety of percents. These values will vary around the true percent of women in the United States (excluding Alaska and Hawaii) who feel they do not get enough time for themselves. No particular sample, however, will necessarily give the true value of this percent. Thus we cannot be sure that 47% is the true percent of women in the United States (excluding Alaska and Hawaii) who feel they do not get enough time for themselves. All we can say is that 47% is likely to be "close" to the true percent.

(c) Suppose we take all possible random samples of 1025 women. In each sample, suppose we determine the percent of women who think they do not get enough time for themselves. For each of these percents, suppose we add and subtract the margin of error for a 95% confidence interval. 95% of the resulting intervals will contain the actual percent of all adult women in the United States (excluding Alaska and Hawaii) who think they do not get enough time for themselves. This is what we mean by "95% confidence." Note that we do not know if any particular interval (such as the 44% to 50% interval in part (a) contains the true value of the percent. 95% only refers to the percent of the intervals produced by all samples that will contain the true percent.

Exercise 5.7

(a) Here are two histograms, with different interval widths.

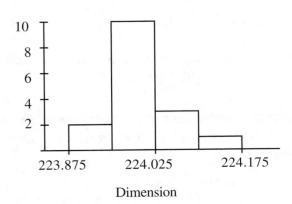

The general shape is roughly symmetric and bell shaped (particularly in the histogram on the left), but with slight skewness to the right.

(b) We find that the mean of the measurements is $\bar{x} = 224.00194$. Since $\sigma = .060$, we find that a 95% confidence interval for the process mean μ is

$$\bar{x} \pm z^* \frac{\sigma}{\sqrt{n}} = 224.00194 \pm 1.96 \frac{.06}{\sqrt{16}} = 224.00194 \pm .0294$$

or (223.97254, 224.03134).

Exercise 5.11

(a) We are told that the sample mean $\bar{x}$ is 275, the standard deviation σ is 60, and the sample size n is 1077. Thus a 95% confidence interval for the mean score μ in the population of all young women is

$$\bar{x} \pm z^* \frac{\sigma}{\sqrt{n}} = 275 \pm 1.96 \frac{60}{\sqrt{1077}} = 275 \pm 3.58$$

or (271.42, 278.58).

(b) In computing 90% and 99% confidence intervals we replace the $z^* = 1.96$ in part (a) by 1.645 and 2.576, respectively. This yields

90% confidence interval for μ is $275 \pm 1.645 \frac{60}{\sqrt{1077}} = 275 \pm 3.01 = (271.99, 278.01)$

99% confidence interval for μ is $275 \pm 2.576 \frac{60}{\sqrt{1077}} = 275 \pm 4.71 = (270.28, 279.71)$.

(c) The margins of error, calculated in parts (a) and (b), are

90% confidence interval for μ: ±3.01
95% confidence interval for μ: ±3.58
99% confidence interval for μ: ±4.71

We see that as the confidence level increases the margin of error also increases.

Exercise 5.15

We recall that the standard deviation for the dimension in Exercise 5.7 is $\sigma = .060$. Thus to estimate the mean μ with a margin of error $m = 0.020$ with 95% confidence we need a sample size of

$$n = (z^*s/m)^2 = ((1.96)(.060)/.020)^2 = (5.88)^2 = 34.57.$$

Since this is not a whole number, we round up to $n = 35$.

Exercise 5.17

(a) Suppose we listed all possible samples we could obtain by the method used to get the one actually taken. For each, suppose we calculated the percent of the sample that intended to vote for Carter and then attached the appropriate margin of error (the ± 2% for the sample in the article) to each of these percents. 95% of the resulting intervals would contain the true percent of people that intended to vote for Carter. Notice that the 95% tells us the percent of samples collected identically to the one actually taken which would contain the true percent of voters favoring Carter. Any particular sample (such as the one actually taken of 51% ± 2%) either does or does not contain the true percent.

(b) The 95% confidence interval for the true percent of voters favoring Carter was 51% ± 2% or (49%, 53%). This interval includes values below 50%, which does not rule out the possibility that less than 50% favor Carter. As a result, the polling organization had to say that the election was too close to call (using a 95% confidence level).

(c) The true percent of the voters that favor Carter is not known, but is a fixed value. This fixed value is either larger than 50% or it is not larger than 50%. There is no probability involved. Thus the politician's question does not really make sense. Most likely, the politician is confusing a lack of knowledge, or uncertainty, about the true value with the notion of probability or chance.

Remember, probabilities in polls refer to the method of selecting a sample. In particular, they refer to whether the method used to select the sample will produce results within some given margin of error of the true value. Probabilities do not refer to the true value itself. Since the true value is unknown (that is why we are taking the sample in the first place), there is always a temptation to think of probabilities as referring to our lack of knowledge about this true value. In this text, avoid this temptation.

Note: There are statisticians that define probability in terms of our lack of knowledge or uncertainty about unknown parameters of a population. This is not the view adopted by most statisticians. If you take additional statistics courses, however, you may run across this alternate view. It is sometimes called the Bayesian or subjective view of probability.

Exercise 5.21

(a) We would assume that the authors of the study wanted to draw conclusions about the population of all adult American consumers. Since the sample was drawn from the Indianapolis phone directory, we can only be certain that they can draw conclusions about the population of all people listed in this phone directory.

(b) We use the formula $\bar{x} \pm z^* \dfrac{\sigma}{\sqrt{n}}$, but with the sample standard deviations, s, in place of σ. Here the sample size is $n = 201$ and since we want 95% confidence intervals, $z^* = 1.96$. The confidence intervals are

Food stores: $18.67 \pm 1.96 \left(\dfrac{24.95}{\sqrt{201}} \right) = 18.67 \pm 3.45 = (15.22, 22.12)$

Mass merchandisers: $32.38 \pm 1.96 \left(\dfrac{33.37}{\sqrt{201}} \right) = 32.38 \pm 4.61 = (27.77, 36.99)$

Pharmacies: $48.60 \pm 1.96 \left(\dfrac{35.62}{\sqrt{201}} \right) = 48.60 \pm 4.92 = (43.68, 53.52)$

(c) None of the 95% confidence intervals overlap with that for pharmacies. In fact, none are even close to overlapping with the interval for pharmacies. This seems reasonably strong evidence that consumers *from the population of all people listed in the Indianapolis phone directory* think that pharmacies offer higher performance than other types of stores. Whether this is strong evidence that *all consumers* think that pharmacies offer higher performance than other types of stores is less clear. This involves deciding to what extent the population of all people listed in the Indianapolis phone directory are representative of all consumers in the U.S. Without further information, we would be reluctant to extend our conclusions, or extrapolate, to the population of all U.S. consumers.

Exercise 5.23

From Exercise 5.22, we use $s = \$8721$ as the population standard deviation σ. The margin of error is to be $m = \$1000$. For 99% confidence we use $z^* = 2.576$. We therefore need a sample size of

$$n = (z^*s/m)^2 = ((2.576)(8721)/(1000))^2 = (22.465296)^2 = 504.7$$

We round this up to the nearest integer, so the required sample size is $n = 505$.

SECTION 5.2

SECTION OVERVIEW

Confidence intervals are one half of the inference duo. The other half is **tests of significance**. We will use these tests when we are trying to evaluate a theory or hypothesis about the population. Significance tests compare two hypotheses about the population: the **null hypothesis** and the **alternative hypothesis**. The test measures the strength of the evidence against the null hypothesis.

Like confidence intervals, significance tests have a probability statement associated with them. The **P-value** states how unlikely a result at least as extreme (as high or as low) as the one we observed would be if the null hypothesis was true. Small P-values are called **statistically significant** and support the alternative hypothesis. Use of any significance test requires use of your best judgment to decide if the results are meaningful.

KEY CONCEPTS

Stating the hypotheses

Both of the hypotheses need to be written in terms of the population parameter we are interested in.

H_0 stands for the null hypothesis. Usually the null hypothesis is the hypothesis of "no effect." This is usually written $\mu = \mu_0$, where μ_0 is the value the population parameter would have under the assumption that no effect is present.

H_a stands for the alternative hypothesis. The alternative hypothesis usually says that there is a real effect present. Generally, it is what we believe to be true or what we are trying to prove true. H_a can be one-sided or two-sided. Two-sided means we care if the parameter is either higher **or** lower than the value under the null hypothesis. One-sided says we are only interested if the parameter differs from the value under the null hypothesis in one direction. (*Hint:* a two-sided alternative uses the not equal to sign, $\neq$. A one-sided alternative uses greater than or less than, $<>$.)

P-values and significance

P-values show how surprising the value of the statistic that we measured is. *P*-values are probabilities, the probability of getting a value as big (or as small) as the one we found. In this chapter we calculate these probabilities using the normal probability methods we met in Chapter 1. Small values support the alternative hypothesis. Large values favor the null.

We call test results significant if the *P*-value is smaller than a predetermined value called the significance level, usually written as α. The most common pre-set level for α is 0.05. Often, when we have a fixed significance level, we compare the *P*-value with it; if the *P*-value is lower then we say the result is significant.

A test about the population mean

This test follows the same pattern as all significance tests.

1. State the alternative hypothesis. That is, write out the effect you are looking for in terms of the population parameter.

2. State the null hypothesis. Write out in terms of the population parameter that there is no effect.

3. Calculate the statistic, now called the test statistic, from the data.

4. Find the *P*-value.

For any test about the population mean, the test statistic is the standardized sample mean, $z = \dfrac{\bar{x} - \mu_0}{\sigma / \sqrt{n}}$, where μ_0 is the value used in the statement of the null hypothesis. How to calculate the *P*-value depends on the way the alternative hypothesis is written.

Find the probability $P(Z \geq z)$ if H_a is $\mu > \mu_0$
Find the probability $P(Z \leq z)$ if H_a is $\mu < \mu_0$
Find the probability $2P(Z \geq |z|)$ if H_a is $\mu \neq \mu_0$

SOLUTIONS TO SELECTED TEXT EXERCISES

Exercise 5.27

(a) If we can assume that the distribution of the percent of spending devoted to housing among all Cleveland households is not too non-normal, then the sampling distribution of the mean percent $\bar{x}$ that the sample of 40 households spends on housing will be approximately normal by the central limit theorem. If the null hypothesis is true and the population standard deviation is $\sigma = 9.6\%$, then the sampling distribution of $\bar{x}$ is approximately normal with mean $\mu = 31\%$ and standard deviation $\dfrac{\sigma}{\sqrt{n}} = \dfrac{9.6\%}{\sqrt{40}} = 1.52\%$.

A sketch of this distribution is

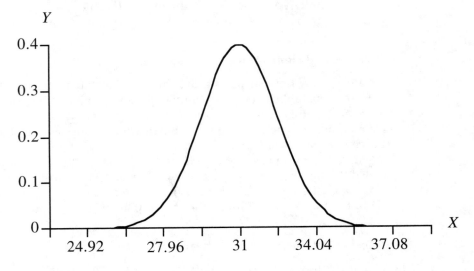

(b) Below is the graph of the sampling distribution with $\bar{x} = 30.2\%$ and $\bar{x} = 27.6\%$ marked on it.

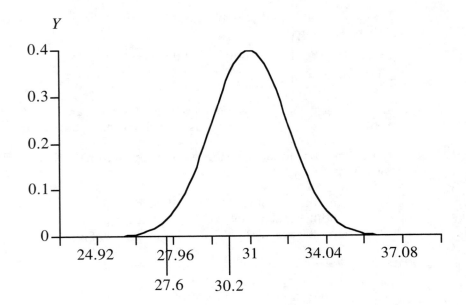

If we examine the sampling distribution of $\bar{x}$, we see that a result as small as $\bar{x}$ = 30.2% is not surprising if the null hypothesis is true. 30.2% is fairly close to the center of the sampling distribution and is less than one standard deviation below μ = 31%. It is, therefore, not compelling evidence that the average Cleveland spending on housing is less than 31%. $\bar{x}$ = 27.6%, however, is more than two standard deviations below μ = 31%. A result as small as $\bar{x}$ = 27.6% is unusual if the null hypothesis is true and so is good evidence that the average Cleveland spending on housing is less than 31%.

(c) The area under the sampling distribution that gives the P-value for the result $\bar{x}$ = 30.2% is the area to the left of 30.2%. This is because we are looking for evidence that the average spending is less than 31%. This area is indicated in the graph below.

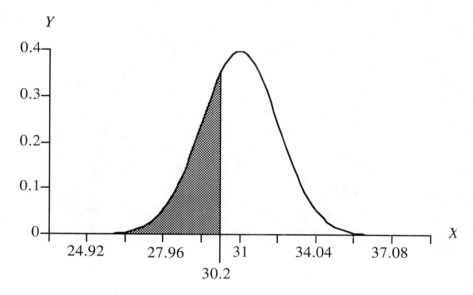

Exercise 5.31

Evidently, we wish to see if this year's data indicate that the average time μ to respond to trouble calls from business customers who had purchased service contracts *differs* (in either direction) from 2.6 hours (last year's average and hence indicative of no change). This suggests that the null hypothesis should be H_0: μ = 2.6 and the alternative hypothesis should be H_a: $\mu \neq 2.6$.

Exercise 5.33

The P-value for both $\bar{x}$ = 30.2% and $\bar{x}$ = 27.6% are the areas under the sampling distribution (see graphs in Exercise 5.27) of $\bar{x}$ under the null hypothesis to the left of these values. Since the sampling distribution of $\bar{x}$ under the null hypothesis is approximately normal with mean 31% and standard deviation 1.52%, the P-values are

P-value for 30.2% = $P(\bar{x} \leq 30.2)$ = $P\left(\dfrac{\bar{x} - 31}{1.52} \leq \dfrac{30.2 - 31}{1.52}\right)$ = $P(Z \leq -0.53) = .2981$

P-value for 27.6% = $P(\bar{x} \leq 27.6)$ = $P\left(\dfrac{\bar{x} - 31}{1.52} \leq \dfrac{27.6 - 31}{1.52}\right)$ = $P(Z \leq -2.24) = .0125$

We see that a value as small as 30.2% is not particularly unlikely under the null hypothesis while a value as small as 27.6% is.

Exercise 5.35

(a) If we assume that the distribution of increase in cash compensation of CEOs is not too non-normal, the sampling distribution of $\bar{x}$ will be approximately normal by the central limit theorem. The sampling distribution will have mean $\mu = 0$ when H_0 is true and standard deviation $\dfrac{\sigma}{\sqrt{n}} = \dfrac{55\%}{\sqrt{104}} = 5.39\%$. A sketch of this sampling with the P-value for the observed outcome $\bar{x} = 6.9$ is given below. Notice that the P-value is the area to the right of 6.9 since the alternative hypothesis is that $\mu > 0$. Thus large values of the sample mean are evidence against H_0.

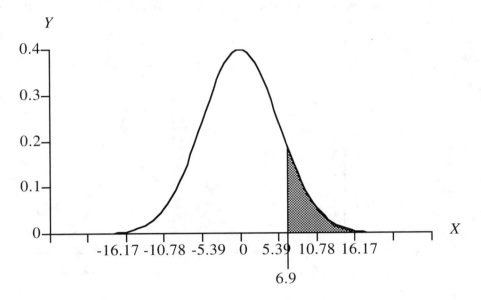

(b) The P-value for the observed outcome $\bar{x} = 6.9$ is

$$\text{P-value for } 6.9 = P(\bar{x} \geq 6.9) = P\left(\dfrac{\bar{x} - 0}{5.39} \geq \dfrac{6.9 - 0}{5.39}\right) = P(Z \geq 1.28)$$

$$= 1 - P(Z < 1.28) = 1 - .8997 = .1003$$

(c) Since the P-value is larger than $\alpha = .05$, the result is not significant at $\alpha = .05$. The P-value of .1003 has the following interpretation. If H_0 is true, in 10.03% of all samples of 104 CEOs we would obtain a value of $\bar{x}$ as large as 6.9. 10.03% is small, but not unusually small, so we might say that the study gives weak evidence that the mean compensation of all CEOs went up.

Exercise 5.39

(a) We wish to determine whether the mean contents are less than the advertised 300 ml. This suggests that we take H_0 as $\mu = 300$ ml, since this will be the case if there is no departure from the advertised value, and we take H_a as $\mu < 300$ ml, since this will be the case if the machinery is underfilling the bottles. Our hypotheses are therefore

$$H_0: \mu = 300 \text{ ml}$$
$$H_a: \mu < 300 \text{ ml}$$

(b) The mean of the $n = 6$ measurements is $\bar{x} = 299.03$. Since the distribution of the contents of bottles is normal with standard deviation $\sigma = 3$ ml, the test statistic is

$$z = \frac{\bar{x} - \mu_0}{\sigma / \sqrt{n}} = \frac{299.03 - 300}{3 / \sqrt{6}} = -0.79$$

(c) The P-value is the probability that a standard normal variable Z takes a value less than or equal to -0.79. This is indicated in the graph below.

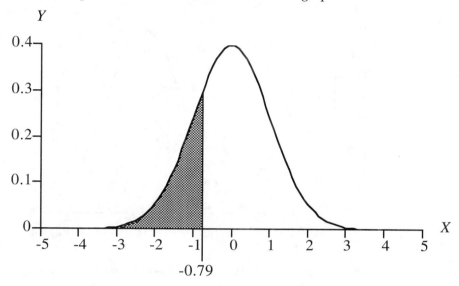

We find from Table A,

$$P = P(Z \leq -0.79) = .2148$$

This P-value is not unusually small. The sample does not provide strong evidence that the mean contents of cola bottles is less than the advertised 300 ml.

Exercise 5.41

(a) Here we have a one-sided alternative. Therefore at the 5% level we use the upper 5% critical value in Table C, namely $z^* = 1.645$. Since the test statistic $z = 2.42$ is larger than $z^* = 1.645$, the result is significant at the 5% level.

(b) Since we have a one-sided alternative, at the 1% level we use the upper 1% critical value in Table C, namely $z^* = 2.326$. The test statistic $z = 2.42$ is larger than $z^* = 2.326$, so the result is significant at the 1% level.

Exercise 5.43

(a) If the null hypothesis is true, the sampling distribution of the mean JDS score $\bar{x}$ is normal with mean 0 (the value of the mean under the null hypothesis) and standard deviation $\frac{\sigma}{\sqrt{n}} = \frac{.60}{\sqrt{28}} = 0.11$. A graph of this density curve is given below.

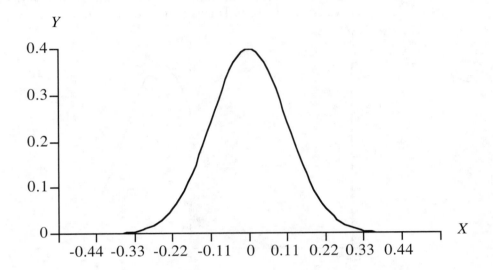

(b) Our graph with the two values of $\bar{x}$ marked on it is reproduced below.

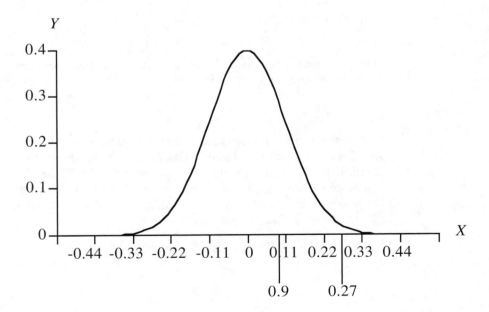

Looking at the graph we see that the value of .09 is near the center of the sampling distribution (within 1 standard deviation of the mean) while .27 is much farther out in the tail (more than 2 standard deviations away from the mean). A value as extreme as .27 is much more unlikely under this sampling distribution than a value as extreme as .09. The value of .27 would be good evidence that H_0 is not true. The value of .09 would not.

(c) The area under the curve below is the *P*-value of the result $\bar{x} = .09$. Recall that we have a two-sided alternative and so must take into account values as far away from $\mu = 0$ both above and below 0.

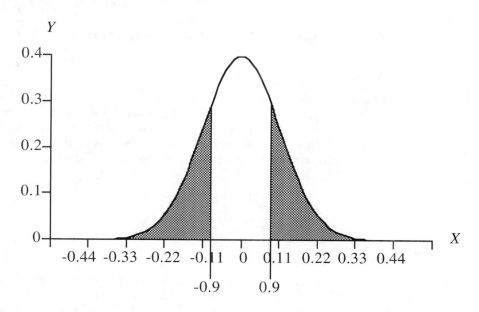

The *P*-value is

$$P\text{-value} = 2P(\bar{x} \geq .09) = 2\ P\left(\frac{\bar{x} - 0}{.11} \geq \frac{.09 - 0}{.11}\right) = 2P(Z \geq .82)$$

$$= 2(1 - P(Z < .82)) = 2(1 - .7939) = .4122$$

(d) The area under the curve below is the *P*-value of the result $\bar{x} = .27$. Recall that we have a two-sided alternative and so must take into account values as far away from $\mu = 0$ both above.and below 0.

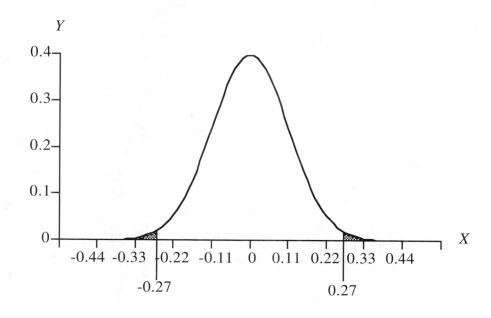

The *P*-value is

$$P\text{-value} = 2P(\overline{x} \geq .27) = P\left(\frac{\overline{x} - 0}{.11} \geq \frac{.27 - 0}{.11}\right) = 2P(Z \geq 2.45)$$
$$= 2(1 - P(Z < 2.45)) = 2(1 - .9929) = .0142$$

We see that, indeed, $\overline{x} = .27$ is much stronger evidence against H_0 than $\overline{x} = .09$.

Exercise 5.45

If the noise has no effect, then there should be no change in the mean time of $\mu = 18$ seconds that it takes to complete the maze. This suggests that we take the null hypothesis of no effect to be H_0: $\mu = 18$ seconds. If the loud noise causes the mice to complete the maze faster, than the mean time μ should decrease. This suggests that we take the alternative hypothesis to be H_a: $\mu < 18$ seconds. Summarizing, our hypotheses are

$$H_0: \mu = 18 \text{ seconds}$$
$$H_a: \mu < 18 \text{ seconds}$$

Exercise 5.51

Before we can calculate the *P*-value, we must understand what hypotheses are being tested. In this study we are interested in whether the mean increase is positive in the average market share value μ of those with pioneer share advantages over those without this benefit. If there is no advantage (no effect) μ should be 0. If there is an advantage, μ should be > 0. Thus the null and alternative hypotheses tested are

$$H_0: \mu = 0$$
$$H_a: \mu > 0$$

With this in mind, the *P*-value for $z = 1.13$ is (we assume that the test statistic is based on the average of 1209 share values so that by the central limit theorem we can approximate the sampling distribution of this average by a normal distribution) indicated by the shaded area in the graph below.

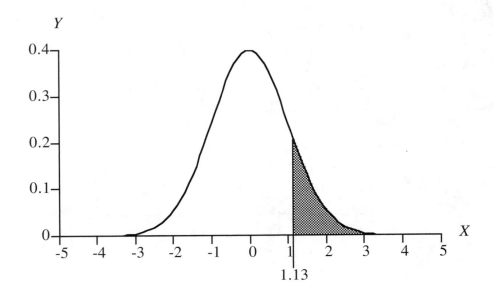

The shaded area has value

$$P = P(z \geq 1.13) = 1 - P(z < 1.13) = 1 - .8708 = .1292$$

In this case, "not statistically significant" means that the probability of obtaining a sample mean difference as large as 2 points higher than that of pioneers without this benefit under the assumption that there is no real difference, is not small enough that it would cause us to doubt that there is no real difference. Even though patents and trade secrets contributed two percentage points to the average market share of the sampled manufacturers, such an increase can be reasonably attributed to chance. If there is no real benefit of patents and trade secrets, there is a 12.92% chance that further samples of size 1209 would also yield a difference as large as 2 percentage points.

SECTION 5.3

SECTION OVERVIEW

Section 5.3 is probably the most important section in this chapter. No matter how good we are at using the recipes to conduct a significance test or find a confidence interval, if our use is wrong then the results are useless. Significance tests are often used to guide decisions and so we need to know the best ways to use the information these tests provide. Here are some tips to remember about significance tests.

- Report the actual *P*-value. This will let users of your work decide for themselves if the result is significant.

- A result that is statistically significant is not necessarily of practical significance.

- Use all the information you have to be sure the use of the test was OK. Outliers are especially dangerous. Good tests cannot correct poorly collected data.

- Is a significance test the best tool to use in this case? A confidence interval may be more appropriate for judging the importance of an effect.

- Do not perform many tests at once. The *P*-values will no longer be valid and some of the tests may show statistical significance just by chance.

SOLUTIONS TO SELECTED TEXT EXERCISES

Exercise 5.55

(a) Under the null hypothesis, $\mu_0 = 475$. The population standard deviation is $\sigma = 100$ and the sample size is $n = 100$. The test statistic is therefore

$$z = \frac{\bar{x} - \mu_0}{\sigma / \sqrt{n}} = \frac{491.4 - 475}{100 / \sqrt{100}} = 1.64$$

The upper 5% critical value is $z^* = 1.645$ and we reject H_0: $\mu = 475$ against H_a: $\mu > 475$ if $z \geq z^* = 1.645$. Since $z = 1.64 < z^* = 1.645$ we do not reject H_0 at the 5% level. $\bar{x} = 491.4$ is not significant at the 5% level.

(b) The test statistic here is

$$z = \frac{\bar{x} - \mu_0}{\sigma / \sqrt{n}} = \frac{491.5 - 475}{100 / \sqrt{100}} = 1.65$$

Now $z = 1.65 > z^* = 1.645$, so we reject H_0 at the 5% level. $\bar{x} = 491.5$ is significant at the 5% level! An increase of 0.1 in the value of $\bar{x}$ has changed our result from not significant at the 5% level in part (a) to significant in part (b).

Exercise 5.57

(a) A 99% confidence interval for the mean SATM score μ based on the results of coaching $n = 100$ students is (using the values of $\bar{x} = 478$ and $\sigma = 100$ from Exercise 5.56 and $z^* = 2.576$ for 99% a confidence level)

$$\bar{x} \pm z^* \frac{\sigma}{\sqrt{n}} = 478 \pm 2.576 \frac{100}{\sqrt{100}} = 478 \pm 25.76 = (452.24, 503.76)$$

(b) Now, a 99% confidence interval for the mean SATM score μ based on the results of coaching $n = 1000$ students is

$$\bar{x} \pm z^* \frac{\sigma}{\sqrt{n}} = 478 \pm 2.576 \frac{100}{\sqrt{1000}} = 478 \pm 8.15 = (469.85, 486.15)$$

(c) Now, a 99% confidence interval for the mean SATM score μ based on the results of coaching $n = 10,000$ students is

$$\bar{x} \pm z^* \frac{\sigma}{\sqrt{n}} = 478 \pm 2.576 \frac{100}{\sqrt{10,000}} = 478 \pm 2.58 = (475.42, 480.58)$$

Notice that for larger samples the margin of error gets smaller. Eventually (at $n = 10,000$) the margin of error is small enough so that at a 99% confidence level, a mean of $\bar{x} = 478$ is different from 475, although by a small amount.

Exercise 5.59

(a) A P-value of .01 means that the probability a subject would do so well when merely guessing is only .01. Among 500 subjects, all of whom are merely guessing, we would therefore expect 1%, or 5, of them to do as well. Thus in 500 tests it is not unusual to see 4 results with P-values on the order of .01, even if all are guessing and none have ESP.

(b) These four subjects only should be re-tested with a new, well designed test. If all four again have low P-values (say, below .01 or .05), we have real evidence that they are not merely guessing. In fact, if any one of the subjects has a very low P-value (say, below .01) this would also be reasonably compelling evidence that the individual is not merely guessing. A single P-value on the order of .10, however, would not be particularly convincing.

SECTION 5.4

(optional)

SECTION OVERVIEW

In this section we look at using significance tests to make decisions. Here we are no longer concerned about collecting evidence against H_0, but we look at the hypotheses as two choices we must select between. In making a decision there are two kinds of possible errors: rejecting H_0 when H_0 is true (**Type I error**) and choosing H_0 when H_a is true (**Type II error**). Each type of error has a probability associated with it. We hope to minimize the probabilities of both types by the way our tests are set up. The significance level, α is the probability of a Type I error. We can calculate the probability of a Type II error using the same probability calculations we have seen all along. The **power** of a test is $1-P$(Type II error). Power is another measure of the sensitivity of the test.

SOLUTIONS TO SELECTED TEXT EXERCISES

Exercise 5.63

(a) The two hypotheses are

H_0: the patient has no medical problem
H_a: the patient has a medical problem

One possible error is to decide

H_a: the patient has a medical problem

when, in fact, the patient does not really have a medical problem. This is a Type I error and in this setting could be called a false positive. The other type of error is to decide

H_0: the patient has no medical problem

when, in fact, the patient does have a problem. This is a Type II error and in this setting could be called a false negative.

Note that one could also switch H_0 and H_a as specified above. While we usually take the hypothesis of "no effect" as H_0, when testing is viewed as a decision problem one can interchange H_0 and H_a. One simply has to be careful to keep track of what one is calling a Type I and Type II error. Interchanging H_0 and H_a will "switch "our answers above and in what follows.

(b) We would choose to decrease the error probability for a Type II error, or the false negative probability. Failure to detect a problem (particularly a major problem) when one is present could result in serious consequences (such as death). While a false positive can also have serious consequences (painful or expensive treatment that is not necessary), it is not likely to lead to the kinds of consequences that a false negative could produce. For example, consider the consequences of failure to detect a heart attack, the presence of AIDS, or the presence of cancer. Note, there are cases where some might argue that a false positive would be a more serious error than a false negative. For example, a false positive in a test for downs syndrome or a birth defect in an unborn baby might lead parents to consider an abortion. Some would consider this a much more serious error than to give birth to a child with a birth defect.

Exercise 5.65

(a) If $\mu = 0$, the sampling distribution of $\bar{x}$ is normal with mean 0 and standard deviation $\frac{\sigma}{\sqrt{n}} = \frac{1}{\sqrt{9}} = 0.33$. Thus the probability of a Type I error is

$$P(\bar{x} > 0) = P\left(\frac{\bar{x} - 0}{0.33} > \frac{0 - 0}{0.33}\right) = P(Z > 0) = 0.5$$

(b) If $\mu = 0.3$, the sampling distribution of $\bar{x}$ is normal with mean 0.3 and standard deviation $\frac{\sigma}{\sqrt{n}} = \frac{1}{\sqrt{9}} = 0.33$. We accept H_0 if $\bar{x} \leq 0$. Thus the probability of a Type II error when $\mu = 0.3$ is

$$P(\bar{x} \leq 0) = P\left(\frac{\bar{x} - 0.3}{0.33} \leq \frac{0 - 0.3}{0.33}\right) = P(Z \leq -0.91) = .1814$$

(c) If $\mu = 1$, the sampling distribution of $\bar{x}$ is normal with mean 1 and standard deviation $\frac{\sigma}{\sqrt{n}} = \frac{1}{\sqrt{9}} = 0.33$. We accept H_0 if $\bar{x} \leq 0$. Thus the probability of a Type II error when $\mu = 1$ is

$$P(\bar{x} \leq 0) = P\left(\frac{\bar{x} - 1}{0.33} \leq \frac{0 - 1}{0.33}\right) = P(Z \leq -3.0) = .0013$$

Exercise 5.67

We begin by rewriting the rule for rejecting H_0 in terms of $\bar{x}$. The rule is to reject H_0 if $z \leq -1.645$, or

$$\frac{\bar{x} - 300}{3/\sqrt{6}} = \frac{\bar{x} - 300}{1.22} \leq -1.645$$

Rewriting this in terms of $\bar{x}$ gives the following rule for rejection. Reject H_0 if

$$\bar{x} \leq (1.22)(-1.645) + 300 = 297.99$$

(a) If $\mu = 299$, the sampling distribution of $\bar{x}$ is normal with mean 299 and standard deviation $\frac{3}{\sqrt{6}} = 1.22$. The power is the probability of rejecting H_0 when the particular alternative $\mu = 299$ is true. This probability is

$$\text{Power} = P(\bar{x} \leq 297.99) = P\left(\frac{\bar{x} - 299}{1.22} \leq \frac{297.99 - 299}{1.22}\right) = P(Z \leq -0.83) = .2033$$

(b) If $\mu = 295$, the sampling distribution of $\bar{x}$ is normal with mean 295 and standard deviation $\dfrac{3}{\sqrt{6}} = 1.22$. The power is the probability of rejecting H_0 when the particular alternative $\mu = 295$ is true. This probability is

$$\text{Power} = P(\bar{x} \le 297.99) = P\left(\frac{\bar{x} - 295}{1.22} \le \frac{297.99 - 295}{1.22} \right) = P(Z \le 2.45) = .9929$$

(c) The power will be higher than in part (b). We notice in parts (a) and (b) that as μ decreased, the power increased. Thus we would suspect that the power will be higher when μ decreases further to 290. A more precise argument is the following. If $\mu = 290$, $\bar{x}$ is likely to be close to 290 (within a standard deviation or two of 290). 297.99 is several multiples of the standard deviation of 1.22 above 290. Thus it is almost certain that $\bar{x}$ will be below 297.99 and more likely to be below 297.99 than if μ was 295 (which is not as far below 297.99 as 290 is). This means the power will be very close to 1 and higher than in part (b).

Exercise 5.71

Recall that the test in Exercise 5.67 used a significance level of 5% to test the hypotheses

$$H_0: \mu = 300$$
$$H_a: \mu < 300$$

The probability of a Type I error is the same as the significance level and hence is .05. The probability of a Type II error at the alternative $\mu = 295$ is the probability of accepting the null hypothesis $H_0: \mu = 300$. One minus this probability is the probability of (correctly) rejecting the null hypothesis $H_0: \mu = 300$ when the alternative $\mu = 295$ is true. This last probability is the power at the alternative $\mu = 295$. We found this to be .9929 in part (b) of problem 5.67. One minus this value is thus the Type II error. Hence the Type II error is $1 - .9929 = .0071$.

SOLUTIONS TO SELECTED TEXT REVIEW EXERCISES

Exercise 5.77

(a) If there is no difference between the mean score μ of students in the suburban district and the national mean, μ will be 32. This condition of "no difference" should be our null hypothesis. The researcher believes that there may be a difference and that the mean μ is higher than the national mean of 32. This should be our alternative hypothesis. Thus the hypotheses to be tested are

$$H_0: \mu = 32$$
$$H_a: \mu > 32$$

(b) The mean $\bar{x}$ of the SRS of 44 scores is $\bar{x} = 35.09$. Since the alternative hypothesis is $H_a: \mu > 32$, the P-value is the probability of observing a sample mean as large or larger than 35.09 if H_0 is true. The sampling distribution of $\bar{x}$ if H_0 is true is normal with mean 32 and standard deviation $\dfrac{\sigma}{\sqrt{n}} = \dfrac{11}{\sqrt{44}} = 1.66$. Thus the P-value is

$$P\text{-value} = P(\bar{x} \geq 35.09) = P\left(\frac{\bar{x} - 32}{1.66} \geq \frac{35.09 - 32}{1.66} \right) = P(Z \geq 1.86)$$
$$= 1 - P(Z < 1.86) = 1 - .9686 = .0314$$

There is only a 3.14% chance of observing a sample mean as large as 35.09 (in an SRS of 44 scores) if the school district mean is actually 32. This probability is fairly small and is unlikely under the hypothesis that the district mean is 32, the same as the national mean. We therefore have reasonably good evidence that the district mean is larger than the national mean.

Exercise 5.79

(a) Increasing the sample size reduces the margin of error of a level C confidence interval.

(b) If H_0 is false, as the sample size increases the value of the sample estimate (such as the sample mean) is more and more likely to be close to the true value of the population parameter of interest. This implies that the sample estimate will be less consistent with the null hypothesis as the sample size increases and hence we will be less and less likely to see a value as extreme as that observed if the null hypothesis is true. This, in turn, implies that the P-value will decrease as the sample size increases.

(c) Suppose that a particular value of the alternative hypothesis is true. As the sample size increases, the sample estimate of the parameter of interest is more and more likely to be close to this particular value of the parameter under H_a. For a fixed level α test, this means that the test statistic will take on values less and less likely under H_0, i.e., larger and larger values. This, in turn, means that the probability of rejecting H_0 when the particular value of the alternative is true, will increase as the sample size increases. Since the power is precisely this probability, we see that the power of a fixed level α test, when H_a, the alternative hypothesis, and all facts about the population remain unchanged, will increase as the sample size increases.

Exercise 5.81

This is not a correct explanation. Remember, the null hypothesis is either true or false. There is no probability involved. Probability enters when we talk about the possible values we will obtain from a sample which will be used to test the null hypothesis. Probability refers to the chance our sample will give us reliable or misleading information. A correct explanation of the P-value of .03 is that if the null hypothesis is true, there is only a probability of 3% that we would observe a sample result as inconsistent with the null hypothesis as was actually observed, by chance. In other words, P-values tell us the probability that our sample results are due to chance (assuming the null hypothesis is true) as opposed to being the result of a real effect.

CASE STUDY

If you have understood the notions of confidence, *P*-values, and significance levels you are among the statistical elite. These are difficult concepts and are not understood by most people. Yet in order to appreciate how statistics are used in making decisions (and recognize what statistics are not saying), you must understand these concepts. Because they are not well understood by most people, many are deeply suspicious of statistical arguments and are easily misled by poorly designed studies and by improper conclusions drawn from even well-designed studies. An example of the confusion that can arise appeared in a newspaper article (*Columbus Dispatch*, July 3, 1994) about the dangers of secondhand smoke. We quote selected portions of this article below and have underlined the "statistical" statements. Read them before proceeding to our discussion and see if you can make sense out of the statements in the article.

> Philip Morris capped off a week-long attack (on studies concerning secondhand smoke) with a three-page ad in 40 Sunday newspapers that charged the Environmental Protection Agency with using flawed science to label secondhand smoke a carcinogen. ...

> Philip Morris is reprinting an article by a media critic that claims the EPA, in labeling secondhand smoke a carcinogen, used invalid studies and skewed statistics, calling a study significant when it had only a 90 percent chance of accuracy instead of the usual 95 percent chance. ...

> Last week, the EPA insisted that 24 of its 30 studies linked secondhand smoke to cancer and nine were statistically significant.

> The statistics are tricky, but using a 90 percent "confidence interval" is OK when scientists are sure a substance won't have a particular effect, said Dr. Ron Davis, editor of the international journal *Tobacco Control*. In other words, no one says secondhand smoke is good. So 90 percent was enough to detect either no effect or a bad one, and was the same level EPA used to label radon and dioxin dangerous.

The underlined statements are a bit puzzling, even to a student of statistics. We are used to expressing statistical significance as a *P*-value or a significance level. Values like 90% or 95% are typically associated with confidence intervals, not statistical significance. Can we decipher what the article is talking about?

We believe the answer is to be found in the relation between confidence intervals and hypothesis (discussed at the end of Section 5.2 in your text), and in the difference between one- and two-sided tests of significance. Take a few minutes to review that material if it seems unfamiliar.

The first underlined comment on the study having a 90% versus a 95% chance of accuracy probably refers to the connection between a 90% or 95% confidence interval and a significance test with a 10% or 5% level of significance, respectively. For example, if a 95% confidence interval for the difference in two means (such as the mean rates of cancer for people regularly exposed to secondhand smoke versus those not regularly exposed to secondhand smoke) does not contain the value 0, we would reject the two sided test of the hypothesis of no difference in the means at the 5% level. The gist of the statement appears, at first glance, to be that the EPA used a significance level of 10% but should have used a significance level of 5%. The implication is that the EPA used a higher level of significance than usual to make it easier to show that secondhand smoke was a carcinogen; thus the charge of biased scientists and skewed statistics.

If we look at the underlined statements in the last paragraph we see that our first impression is not accurate. These statements indicate that the EPA was only looking for a "bad" effect of secondhand smoke. Only evidence in this direction was considered. The EPA was not interested in evidence that secondhand smoke reduced the risk of cancer. This should suggest to you that the EPA used a one-sided alternative rather than a two-sided alternative. The hypotheses were probably

> H_0: The mean rate of cancer for those exposed to secondhand smoke is the same as for those not exposed.

> H_1: The mean rate of cancer for those exposed to secondhand smoke is greater than that for those not exposed.

If a z-test were used to test these hypotheses (see the boxed expression in the section entitled "Tests with fixed level of significance" in the text), the 5% critical value for such a one-sided alternative would be identical with one of the two 10% critical values for the corresponding two-sided alternative. Other tests also have this property. Presumably the test actually used by the EPA was one of these. Recall that a 90% confidence interval can be used to test a two-sided alternative with a 10% critical value. We therefore have a connection between 90% confidence intervals, tests of two-sided alternatives using a 10% critical value, and tests of one-sided alternatives using a 5% critical value. We believe the last paragraph of the article quoted above is referring to this connection. The key point is not that the EPA used too high a critical value (10% versus the standard 5%) but instead that the EPA tested a one-sided alternative at the 5% level rather than a two-sided alternative at the 5% level. We believe the one-sided alternative was sensible. The EPA was not biased or using skewed statistics as claimed by Philip Morris.

Unfortunately, readers of this article who are not trained in statistics will not recognize these points. They are likely to be suspicious of the whole discussion. To the untrained reader, this article may appear to be another example of using statistics to prove whatever you want. After all, both sides are using the same data to reach different conclusions and attack each others arguments.

Intriguing is the underlined comment in the next-to-last paragraph of the article. The EPA claims that 24 of its 30 studies linked secondhand smoke to cancer and nine were statistically significant. If secondhand smoke is dangerous, wouldn't you expect more than 9 out of 30 of the studies to be statistically significant? Presumably there were 15 studies in which the cancer rate was higher among those exposed to secondhand smoke than those that weren't, but the results were not statistically significant (we assume at the 5% level using a one-sided alternative). What sort of evidence is the EPA's statement? While we cannot be sure, one possible explanation is that the cancer rate is really higher among those exposed to secondhand smoke than those that aren't, but that the difference is not great. The studies may not have been large enough to detect the difference. For those readers that studied the material on power, this means that studies had low power for detecting the actual difference. If this explanation is correct, we then must determine if the difference in cancer rates is "practically significant" even if statistically significant. While any increase in the risk of cancer would seem bad, very small differences may not be worth basing major policy decisions on. After all, there are many activities that increase our risk of injury, disease, or death slightly but are not worth regulating. Only by looking at the actual data can we assess whether the observed differences are of practical importance.

Of course, this explanation may be incorrect. Poorly designed studies containing the same or similar lurking variables might also account for the results. To assess this, we would need descriptions of how the studies were actually conducted.

Finally we note that hidden in the article is an interesting side issue. In the text, we recommend using P-values and examining the actual data, rather than using a fixed significance level and attaching too much importance to it. Statistical significance is not necessarily the same as practical significance. Nevertheless, in many studies, as appears to be the case here, a level of 5% continues to serve as a sort of standard between a result being important and unimportant. This is not the best statistical practice, particularly when making important policy decisions.

CHAPTER 6

INFERENCE FOR DISTRIBUTIONS

CHAPTER OVERVIEW

Some researchers wanted to test the effects of nicotine on pregnancy. Since it would be unethical to do tests like this on humans, the researchers used guinea pigs. They had two groups of young guinea pigs whose mothers were either in the treatment group and received nicotine injections or were in the control group and received saline injections. The young guinea pigs later ran through mazes to test their learning capabilities. This is a common situation: you have two groups of subjects, one receives a certain treatment, one is the control group and receives no treatment. Chapter 6 shows us how we can compare the populations.

Chapter 5 introduced us to the reasoning behind the basic inference methods (confidence intervals and significance tests) and showed us how to use them to learn about the population mean. Chapter 6 will once again demonstrate these tools for the mean of a single population (in Section 6.1) and we will also see how to compare the means of two populations (in Section 6.2). Optional Section 6.3 demonstrates a technique to use when comparing the standard deviations of two populations. When we are done with this chapter we will be able to compare the difference between the nicotine-added guinea pigs and the nicotine-free guinea pigs.

SECTION 6.1

SECTION OVERVIEW

This section will give us the details about confidence intervals and tests of significance about the population mean. In Chapter 5 we looked at the basic ideas but we made the assumption that the population standard deviation, σ, was known. This is not a reasonable assumption. It is a very rare setting where the population standard deviation is known. Now we will learn how to use an estimate of the standard deviation. This is called the **standard error** of the statistic. When we use an estimate, the distribution of the z statistic (the one we used in Chapter 5) becomes the *t* **statistic** and no longer has a normal distribution. It has a *t* **distribution**. The *t* distribution looks a lot like the normal distribution, but the spread is a bit larger, there is more area in the tails, and the distribution has an extra defining parameter, the degrees of freedom. We can use the *t* distribution to calculate confidence interval and perform significance tests.

The *t* procedures are very useful in many situations. These procedures are quite **robust**. The presence of outliers, however, makes these tools unusable. It is very important that the conditions for using the *t* statistics are met. Use the methods from Chapters 1 and 2 to check the conditions.

KEY CONCEPTS

When to use the one-sample t statistic

The one-sample *t* statistic, $t = \dfrac{\bar{x} - \mu_o}{s / \sqrt{n}}$, has a *t* distribution with $n{-}1$ degrees of freedom, written $t(k)$. It has the same form as the standardized value, z, except that the population standard deviation is replaced with s, the sample standard deviation.

The one-sample *t* statistic may be used when there is a single SRS of size n from a normal population with unknown mean, μ, or when there is a matched pairs setting. In the matched pairs setting we use the observed differences to create the statistic. Here, the population of differences must have a normal distribution. It is very important that the sample be an SRS.

Sample size notes:

- $n > 15 \rightarrow$ Use *t* procedures only if data are close to normal. Any doubts? Don't use it.

- n is at least 15 $\rightarrow$ It is OK to use the procedures unless there are outliers or strong skewness.

- n is large $\rightarrow$ For sample sizes over 40, it is OK to use the *t* procedures even if the distribution is clearly skewed. Extreme outliers are still not acceptable.

The one-sample t test and confidence intervals

These procedures work the same way as the z procedures we have seen. Instead of looking up critical values and P–values in the normal table, use the values in the t table for the distribution with n–1 degrees of freedom.

The level C confidence interval has the form

$$\bar{x} \pm t^* \frac{s}{\sqrt{n}}$$

where t^* is the $(1 - C)/2$ critical value from the $t(n–1)$ distribution.

Use the t test to test the hypothesis H_0: $\mu = \mu_0$, where μ_0 is a set value you have chosen. (Usually μ_0 is zero.) Compute the t statistic as shown earlier. Then calculate the P–value based on the alternative hypothesis you have chosen.

If H_a is $\mu > \mu_0$, then find the probability $P(T \geq t)$
If H_a is $\mu < \mu_0$, then find the probability $P(T \leq t)$
If H_a is $\mu \neq \mu_0$, then find the probability $2P(T \geq t)$

SOLUTIONS TO SELECTED TEXT EXERCISES

Exercise 6.3

(a) Since the one-sample t statistic is based on a sample of n =15 observations, it has a t distribution with $n - 1 = 14$ degrees of freedom. To determine the value of t^*, we therefore read across the row labeled 14 (in the df column). The entry in this row under the column labeled has value 2.145. This is therefore the upper tail probability of 2.145 for a t distribution with 14 degrees of freedom. In other words, the area to the right of $t^* = 2.145$ is 0.025.

(b) Here the one-sample t statistic is based on a sample of $n = 20$ observations. It therefore has a t distribution with $n - 1 = 19$ degrees of freedom. We want the value t^* such that the area to the left of it under the t distribution with 19 degrees of freedom is 0.75. This implies that the area to the right of t^* is 0.25, i.e. the upper tail probability is 0.25. Looking under this column in Table C, we see that the entry in the 19 degrees of freedom (df) row is 0.688. Therefore $t^* = 0.688$.

Exercise 6.7

(a) The mean $\bar{x}$ is easily calculated by summing the four measurements and dividing the total by 4. We get

$$\bar{x} = 7.0/4 = 1.75$$

To compute the standard error of the mean we first compute the standard deviation s of the four measurements. We find the variance to be $s^2 = 0.01667$, hence the standard deviation is $s = \sqrt{0.01667} = 0.129$. The desired standard error is obtained by dividing s by $\sqrt{n}$. Since the sample size is $n = 4$, we get

$$\text{standard error of the mean} = \frac{s}{\sqrt{n}} = \frac{0.129}{\sqrt{4}} = 0.0645$$

(b) We use the formula $\bar{x} \pm t^* \frac{s}{\sqrt{n}}$. Since the sample size is $n = 4$ and we want a 90% confidence interval, t^* is the upper .05 critical value for the $t(4 - 1) = t(3)$ distribution. Table C gives this as $t^* = 2.353$. Using our results from part (a), our 90% confidence interval is

$$\overline{x} \pm t^* \frac{s}{\sqrt{n}} = 1.75 \pm 2.353(0.0645) = 1.75 \pm 0.15 = (1.60, 1.90)$$

Exercise 6.9

Step 1. Hypotheses. We wish to test the hypotheses

$$H_0: \mu = 1.3$$
$$H_a: \mu > 1.3$$

Step 2. Test statistic. From Exercise 6.7 we know the basic statistics are

$$\overline{x} = 1.75 \text{ and } \frac{s}{\sqrt{n}} = 0.0645$$

so the test statistic is

$$t = \frac{\overline{x} - \mu_0}{s/\sqrt{n}} = \frac{1.75 - 1.3}{0.0645} = 6.98$$

Step 3. P-value. The P-value for $t = 6.98$ is the area to the right of 6.98 under the t distribution curve with $n - 1 = 3$ degrees of freedom. Using Table C, we search the $df = 3$ row for entries that bracket 6.98.

$$df = 3$$

p	.005	.0025
t^*	5.841	7.453

Since the observed t lies between the critical values for .005 and .0025, the P-value lies between .005 and .0025 for this one–sided test. The data do not support the supposition that the mean absolute refractory period for unpoisoned rats is 1.3 milliseconds.

Exercise 6.11

(a) The most important place to use randomization would be in determining which instrument (right thread vs. left thread) a subject uses first. One could flip a coin with, for example, heads meaning the subject uses the right-hand thread knob first, tails meaning the subject uses the left-hand thread first. Alternatively, in order to balance out the number of times each type is used first, one might choose an SRS of 12 of the 25 subjects. These 12 use the right-hand thread knob first. Everyone else uses the left-hand thread knob first.

A second place one might use randomization is in the order in which subjects are tested. Use a table of random digits to determine this order. Label subjects 01 to 25. The first label that appears in the list of random digits (read in groups of two digits) is the first subject measured. The second label that appears, the next subject measured, etc. This randomization is probably less important than the one described in the previous paragraph. It would be important if the order or time at which a subject was tested might have an effect on the measured response. For example, if the study began early in the morning, the first subject might be sluggish if still sleepy. Sluggishness might lead to longer times and perhaps a larger difference in times. Subjects tested later in the day might be more alert.

(b) The parameter μ would be the true mean difference in the time it takes to use the right-hand thread versus the left-hand thread within matched pairs of right-handed subjects in the population of all such users of instruments using such knobs. In other words

$$\mu = \text{mean time to use right-hand thread} - \text{mean time to use left-hand thread}$$

Since the goal of the study is to determine if right-handed people find the right-hand threads easier to use (which might be interpreted to mean that right-handed people would be able to use the right-hand thread knobs more quickly and would imply the value of μ is less than 0, H_0 and H_a would be

$$H_0: \mu = 0 \text{ (the hypothesis of no effect)}$$

$$H_a: \mu < 0 \text{ (right-handed people find the right-hand thread knobs easier to use)}$$

(c) *Test statistic.* From the data given, we compute the value of the right-thread time minus the left-thread time for each subject and then compute the mean and standard error of the differences. The values of these basic statistics are (since $n = 25$).

$$\bar{x} = -13.32 \text{ and } \frac{s}{\sqrt{n}} = \frac{22.936}{\sqrt{25}} = 4.587$$

so the test statistic is

$$t = \frac{\bar{x} - \mu_0}{s/\sqrt{n}} = \frac{-13.32 - 0}{4.587} = -2.90$$

P-value. The P-value for $t = -2.90$ is the area to the left of -2.90 under the t distribution curve with $n - 1 = 24$ degrees of freedom. By the symmetry of the t distribution, this is the same as the area to the right of $+2.90$. Using Table C, we search the $df = 24$ row for entries that bracket 2.90.

$$df = 24$$

p	.005	.0025
t^*	2.797	3.091

Since the observed t lies between the critical values for .005 and .0025, the P-value lies between .005 and .0025. The data provide strong evidence that right-handed people are able to use the right-hand thread knob more quickly than the left-hand thread knob. This might then be taken to mean that right-handed people find the right-hand thread knobs easier to use.

Exercise 6.15

(a) A histogram and boxplot of the data are given below (one could also make a stemplot of the data)

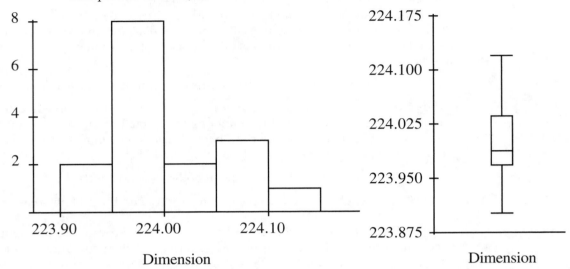

There are no outliers in the data. The data appear a bit skewed to the right, but not so strongly as to threaten the validity of the t procedure given that the sample size is 16 (in the section on the robustness of t procedures, t procedures are safe for samples of size $n \geq 15$ unless there are outliers and/or very strong skewness).

(b) *Step 1. Hypotheses.* Since we are interested in whether the data provide evidence that the mean dimension is not 224 mm. (no direction of the difference is specified), we wish to test the hypotheses

$$H_0: \mu = 224 \text{ mm}$$
$$H_a: \mu \neq 224 \text{ mm}$$

Step 2. Test statistic. From the data we calculate the basic statistics to be

$$\bar{x} = 224.0019 \text{ and } \frac{s}{\sqrt{n}} = \frac{0.0618}{\sqrt{16}} = 0.01545$$

so the test statistic is

$$t = \frac{\bar{x} - \mu_0}{s/\sqrt{n}} = \frac{224.0019 - 224}{0.01545} = 0.123$$

Step 3. P-value. The P-value for $t = 0.123$ is the twice the area to the right of 0.123 under the t distribution curve with $n - 1 = 15$ degrees of freedom. Using Table C, we search the $df = 15$ row for entries that bracket 0.123. We find that 0.123 must lie to the left of (have a P-value greater than)

$$df = 15$$

p	.25
t^*	0.691

since .25 is the largest P-value given in Table C. The P-value is therefore greater than .25 x 2 = .50 for this two-sided test. The data do not provide strong evidence that the mean differs from 224 mm.

Exercise 6.17

(a) The t critical value for $a = .01$ and $n = 50$ is the upper 1% point of the $t(49)$ distribution. Unfortunately Table C does not give upper tail probabilities for $t(49)$, only for $t(40)$ and $t(50)$. We must either take the conservative value of the $t(40)$ distribution, or we can attempt to approximate the value by interpolating between the $t(40)$ and $t(50)$ values. If we use the $t(40)$ value, the critical value is 2.423. If we interpolate (go 9/10 of the way down from 2.423 to 2.403) we get 2.405. For the remainder of this solution, we use the conservative critical value of 2.423 (the interested student should repeat the argument below with 2.405 to see what you would get if you use the critical value of 2.405 obtained by interpolation).

(b) The rule for rejecting H_0 using the critical value 2.423 is to reject if

$$t = \frac{\bar{x} - 0}{s/\sqrt{n}} > 2.423$$

Using $n = 50$ and $s = 108$, this can be written as

$$2.423 < \frac{\bar{x}}{108/\sqrt{50}} = \frac{\bar{x}}{15.2735}$$

or in terms of $\bar{x}$

$$\bar{x} > 2.423 \times 15.2735 = 37.008$$

(Note that in terms of the critical value 2.405, the rule is $\bar{x} > 2.405 \times 15.2735 = 36.733$.)

(c) We need to calculate the probability that $\bar{x} > 37.008$. If we assume that the data are normal with mean $\mu = 100$ and standard deviation $s = 108$, then the sampling distribution of $\bar{x}$ is normal with mean 100 and standard deviation $\dfrac{\sigma}{\sqrt{50}} = \dfrac{108}{\sqrt{50}} = 15.2735$. Thus(note the standard normal table does not give probabilities for values less than –3.49)

$$P(\bar{x} > 37.008) = P(\frac{\bar{x} - 100}{15.2735} > \frac{37.008 - 100}{15.2735}) = P(z > -4.12)$$

$$= 1 - P(z \le -4.12) > 1 - 0.0002 = 0.9998.$$

Thus the power is at least 0.9998, which is very high. The bank can be confident that a test on 50 customers would be sufficient to detect a mean increase of $100. (Note that a similar result will be reached with the critical value of 2.405).

Exercise 6.21

(a) A 95% confidence interval for the mean systolic blood pressure in the population from which the subjects were recruited can be calculated from the data on the 27 members of the placebo group, since these are randomly selected from the 54 subjects. We use the formula for a t–interval, namely $\bar{x} \pm t^* \dfrac{s}{\sqrt{n}}$. In this problem, $\bar{x} = 114.9$, s = 9.3, $n = 27$, hence t^* is the upper $(1 - 95\%)/2 = 2.5\%$ critical value for the $t(26)$ distribution. From Table C we see $t^* = 2.056$. Thus the 95% confidence interval is

$$114.9 \pm 2.056 \frac{9.3}{\sqrt{27}} = 114.9 \pm 3.68 = (111.22, 118.58)$$

(b) For the procedure used in (a), the population from which the subjects were drawn should be such that the distribution of the seated systolic blood pressure in the population is normal. The 27 subjects used for the confidence interval in part (a) should be a random sample from this population. Unfortunately, we do not know if the latter is the case. While 27 subjects were selected at random from the total of 54 subjects in the study, we do not know if the 54 subjects were a random sample from this population.

With a sample of 27 subjects, it is not crucial that the population be normal, as long as the distribution is not strongly skewed and the data contain no outliers. It is important that the 27 subjects can be considered a random sample from the population. If not, we cannot appeal to the central limit theorem to insure that the t-procedure is at least approximately correct even if the data are non normal.

(*Note:* It turns out that since the subjects were divided at random into treatment and control groups, there do exist procedures for comparing the treatment and placebo groups. These are not based on the t distribution, but are valid as long as treatment groups are determined by randomization. However, the conclusions drawn from these procedures apply only to the subjects in the study. To generalize the conclusions to a larger population, we must know that the subjects are a random sample from this larger population.)

Exercise 6.23

(a) SEM must stand for Standard Error of the Mean.

(b) The standard error for the mean, $\frac{s}{\sqrt{n}}$, is 0.01. Since we know $n = 3$, we can determine s for the measurements. In particular, since $\frac{s}{\sqrt{n}} = 0.01$, we have

$$s = 0.01 \sqrt{n} = 0.01 \sqrt{3} = 0.0173$$

(c) We use the formula for a t–interval, namely $\bar{x} \pm t^* \frac{s}{\sqrt{n}}$. In this problem, $\bar{x} = 0.84$, $\frac{s}{\sqrt{n}} = 0.01$, $n = 3$, hence t^* is the upper $(1 - 90\%)/2 = 5\%$ critical value for the $t(2)$ distribution. From Table C we see $t^* = 2.92$. Thus the 90% confidence interval for the mean ATP level in dormant brine shrimp embryos is

$$\bar{x} \pm t^* \frac{s}{\sqrt{n}} = 0.84 \pm 2.92 \times 0.01 = 0.84 \pm 0.0292 = (0.8108, 0.8692)$$

Exercise 6.27

We have the ages of the entire *population* of U.S. presidents. If we simply compute the average of all the ages listed, we will know the value of the population mean (assuming we make no arithmetic mistakes). There is therefore no need for statistical procedures since there is no uncertainty concerning the value of the population mean.

(*Note:* We have occasionally heard statistical procedures when applied to a population, such as in this problem, justified as follows. The ages of the presidents listed is a sample from the population of past, present, and *future* presidents. Statistical procedures thus apply to this hypothetical population. We are usually not impressed by this argument. It is hard to see why anyone would be interested in this hypothetical population. Even if we were interested, the ages of past and present presidents are not really a random sample from this population.)

SECTION 6.2

SECTION OVERVIEW

So far we can estimate the population mean with confidence intervals and test to see if the population mean is or is not equal to a value we have chosen. Section 6.2 addresses the most common type of significance test: the **comparison of two population means**. This is the type of test we would use to compare the difference between the two groups of guinea pigs. In this setting we have two distinct, independent SRS from two populations or two treatments on two samples. The samples are not matched pairs. As before, the populations the samples are taken from must be reasonably normal. Here we use the **two–sample t statistic**.

This statistic involves the difference between the two sample means, so we need to know the sampling distribution of $\bar{x}_1 - \bar{x}_2$. The distribution of $\bar{x}_1 - \bar{x}_2$ is not normal but according to the central limit theorem (see Chapter 4) we can find approximately correct P-values and C-levels. These calculations are easiest to use if sample sizes are equal; this is another issue to consider when planning your experiment.

It is very important that we know whether the data were collected on two independent populations or a single population because the methods for each case are different. In the latter case, you may have matched pairs.

KEY CONCEPTS

Notation notes

The population parameters are written:

For population 1, the variable is called x_1, the mean is μ_1, the standard deviation is σ_1.

For population 2, the variable is called x_2, the mean is μ_2, the standard deviation is σ_2.

The sample statistics are written:

From population 1, sample size is n_1, sample mean is $\bar{x}_1$, and the sample standard deviation is s_1.

From population 2, sample size is n_2, sample mean is $\bar{x}_2$, and the sample standard deviation is s_2.

The hypotheses are written either as H_0: $\mu_1 = \mu_2$ and H_a: $\mu_1 > \mu_2$ or

$$H_0: \mu_1 - \mu_2 = 0 \text{ and } H_a\, \mu_1 - \mu_2 > 0, \text{ whichever you feel is clearer.}$$

When to use the two-sample t statistic

Use the two-sample statistic when there are two independent SRSs. Always check for outliers before using any of the two-sample t procedures. If the two samples are of the same size then the procedures are even more robust. You may use the procedures as long as the population distributions have similar shape, even if the shape isn't normal. If the samples are of different sizes, check to see that the populations are reasonably normal according to the guidelines for the one-sample t procedures. (Use the sum of the two-sample sizes in place of n.)

The two-sample t test and confidence intervals

(This is a recap of Option 2. See the optional sections in the text for details on Option 1.) The two-sample t statistic doesn't have a t distribution, so the P-values and C are conservative.

The confidence interval for $\mu_1 - \mu_2$ is $(\bar{x}_1 - \bar{x}_2) \pm t^* \sqrt{\dfrac{s_1^2}{n_1} + \dfrac{s_2^2}{n_2}}$, where t^* is the upper $(1 - C)/2$ critical value for the $t(k)$ distribution and k is the smaller of $n_1 - 1$ and $n_2 - 1$.

The t test of the hypothesis H_0: $\mu_1 = \mu_2$ uses the statistic

$$t = \frac{\bar{x}_1 - \bar{x}_2}{\sqrt{\dfrac{s_1^2}{n_1} + \dfrac{s_2^2}{n_2}}}.$$

Calculate the P-value using the $t(k)$ distribution where k is the smaller of $n_1 - 1$ and $n_2 - 1$ How you calculate the P-values depends on how the alternative hypothesis is written, like with the one-sample statistic.

SOLUTIONS TO SELECTED TEXT EXERCISES

Exercise 6.29

(a) This example involves two samples, the two separate groups of children. Note we are not told of any matching.

(b) This is an example of matched pairs. There is a pair of observations on each child that are the "matched" pair.

Exercise 6.33

(a) A t of 7.36 is so large that we know that it will have a very small P-value (unless the degrees of freedom are very small, but even with 1 degree of freedom, the P-value would be less than 0.05).

(b) The conservative approach uses the smaller sample size minus 1 for the degrees of freedom. In this case, the smaller sample size is 33, so the conservative approach would use $33 - 1 = 32$ degrees of freedom for the t test.

Exercise 6.35

(a) There are several graphical procedures we could use to present the data. We might make side-by-side boxplots of the control and experimental groups. We might make separate histograms for the control and experimental groups. Histograms are given as follows

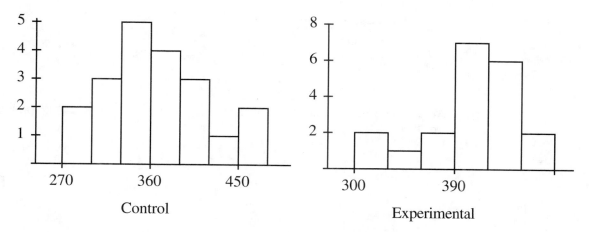

Control   Experimental

We can also make back-to-back stemplots. These are also given below.

```
         2 |27|
         3 |28|
           |29|
           |30|
         6 |31| 8
        91 |32| 6
           |33| 9
        95 |34|
       660 |35|
        60 |36| 1
           |37| 5
        40 |38|
         9 |39| 23
         2 |40| 1367
         0 |41| 0
           |42| 067
         1 |43| 04
           |44|
         5 |45|
         2 |46| 77
           |47| 7
```

Neither plot shows any unusual outliers or extreme skewness. With 20 observations in each group, it is probably safe to use t procedures.

(b) *Step 1.* Let μ_1 represent the mean weight gain on the control feed (feed with normal corn) that would be experienced after 21 days by the population of one-day-old male chicks that are represented by the 20 chicks in the study on the control feed. Let μ_2 represent the mean weight gain on the experimental feed (feed with the new corn) that would be experienced after 21 days by the population of one-day-old male chicks that are represented by the 20 chicks in the study on the experimental feed. Since we wish to determine if chicks on the experimental feed gain weight faster (hence experience a larger weight gain over a fixed period of time) than on the control feed, we would test the hypotheses

$$H_0: \mu_1 = \mu_2$$
$$H_a: \mu_1 < \mu_2$$

Step 2. We must compute the test statistic

$$t = \frac{\bar{x}_1 - \bar{x}_2}{\sqrt{\dfrac{s_1^2}{n_1} + \dfrac{s_2^2}{n_2}}}$$

We calculate $\bar{x}_1 = 366.3$, $s_1^2 = 2581.1684$, $\bar{x}_2 = 402.95$, $s_2^2 = 1825.7342$. Both n_1 and n_2 are 20. Thus

$$t = \frac{366.3 - 402.95}{\sqrt{\dfrac{2581.1684}{20} + \dfrac{1825.7342}{20}}}$$

$$= \frac{-36.65}{\sqrt{220.34513}}$$

$$= -2.469$$

Step 3. We use the conservative procedure with the smaller of $n_1 - 1 = 20 - 1 = 19$ and $n_2 - 1 = 20 - 1 = 19$, or 19 degrees of freedom. The P-value for this test is the area under the $t(19)$ distribution to the left of -2.469. Since the t (19) distribution is symmetric, this is the same as the area under the $t(19)$ distribution to the right of 2.469. From Table C we see

$$df = 19$$

p	.02	.01
t^*	2.205	2.539

Therefore the P-value is between 0.02 and 0.01.

(c) A 95% confidence interval for the mean extra weight gain in chicks fed high-lysine corn (the experimental feed) is given by the formula

$$(\bar{x}_2 - \bar{x}_1) \pm t^* \sqrt{\frac{s_1^2}{n_1} + \frac{s_2^2}{n_2}}$$

where t^* is the upper $.05/2 = .025$ critical value of the $t(19)$ distribution. Table C gives $t^* = 2.093$. From (b) we know $\bar{x}_1 = 366.3$, $s_1^2 = 2581.1684$, $\bar{x}_2 = 402.95$, $s_2^2 = 1825.7342$, so our 95% confidence interval is

$$(\bar{x}_2 - \bar{x}_1) \pm t^* \sqrt{\frac{s_1^2}{n_1} + \frac{s_2^2}{n_2}} = (402.95 - 366.3) \pm (2.093) \sqrt{\frac{2581.1684}{20} + \frac{1825.7342}{20}}$$

$$= 36.65 \pm 31.07 = (6.58, 67.72)$$

Exercise 6.37

We take DDT to be population 1 and the Control to be population 2. The value of t is

$$t = \frac{\bar{x}_1 - \bar{x}_2}{\sqrt{\dfrac{s_1^2}{n_1} + \dfrac{s_2^2}{n_2}}} = \frac{17.6 - 9.499833}{\sqrt{\dfrac{(6.34015)^2}{6} + \dfrac{(1.95006)^2}{6}}} = \frac{8.100167}{\sqrt{7.33337}} = 2.9912$$

thus verifying the computer's results.

Presumably the degrees of freedom were calculated as follows.

$$df = \frac{\left(\dfrac{s_1^2}{n_1} + \dfrac{s_2^2}{n_2}\right)^2}{\dfrac{1}{n_1-1}\left(\dfrac{s_1^2}{n_1}\right)^2 + \dfrac{1}{n_2-1}\left(\dfrac{s_2^2}{n_2}\right)^2}$$

$$= \frac{\left(\dfrac{(6.34015)^2}{6} + \dfrac{(1.95006)^2}{6}\right)^2}{\dfrac{1}{5}\left(\dfrac{(6.34015)^2}{6}\right)^2 + \dfrac{1}{5}\left(\dfrac{(1.95006)^2}{6}\right)^2}$$

$$= \frac{(7.33337)^2}{\dfrac{1}{5}(6.6996)^2 + \dfrac{1}{5}(0.6338)^2}$$

$$= 5.938$$

again verifying the computer's results.

Exercise 6.41

(a) *Step 1. Hypotheses.* This is a two-sample problem. Let the breast-fed group be group 1 and the formula-fed group be group 2. Since we are interested in whether the data provide evidence that the mean hemoglobin levels *differ* for breast-fed and formula-fed babies, we wish to test the hypotheses

$$H_0: \mu_1 = \mu_2$$
$$H_a: \mu_1 \neq \mu_2$$

Step 2. Test statistic. From the basic statistics given, we compute

$$t = \frac{\bar{x}_1 - \bar{x}_2}{\sqrt{\dfrac{s_1^2}{n_1} + \dfrac{s_2^2}{n_2}}} = \frac{13.3 - 12.4}{\sqrt{\dfrac{(1.7)^2}{23} + \dfrac{(1.8)^2}{19}}} = \frac{0.9}{\sqrt{0.296}} = 1.654$$

Step 3. P-value. The P-value for $t = 1.654$ is twice the area to the right of 1.654 under the t distribution curve with the minimum of $n_1 - 1 = 22$ and $n_2 - 1 = 18$ degrees of freedom. This minimum is 18. Using Table C, we search the $df = 18$ row for entries that bracket 1.654. We find

$$df = 18$$

p	.10	.05
t^*	1.330	1.734

The P-value is therefore between $.10 \times 2 = .20$ and $.05 \times 2 = .10$ for this two-sided test. The data do not provide strong evidence that the means differ.

(b) A 95% confidence interval for the mean difference in hemoglobin levels between the two populations of infants is given by the formula

$$\left(\bar{x}_1 - \bar{x}_2\right) \pm t^* \sqrt{\dfrac{s_1^2}{n_1} + \dfrac{s_2^2}{n_2}}$$

where t^* is the upper $.05/2 = .025$ critical value of the $t(k)$ distribution. Here k is the minimum of $n_1 - 1 = 22$ and $n_2 - 1 = 18$, so $k = 18$. Table C thus gives $t^* = 2.101$. Our 95% confidence interval is therefore

$$\left(\bar{x}_1 - \bar{x}_2\right) \pm t^* \sqrt{\frac{s_1^2}{n_1} + \frac{s_2^2}{n_2}} = (13.3 - 12.4) \pm (2.101) \sqrt{\frac{(1.7)^2}{23} + \frac{(1.8)^2}{19}}$$

$$= 0.9 \pm (2.101)(0.544)$$

$$= 0.9 \pm 1.143$$

$$= (-0.243, 2.043)$$

(c) In order to be valid the procedures in (a) and (b) require that we have two SRSs from two distinct populations and the samples be independent. They also require that the populations be either normal or at least that the samples show no extreme skewness and no outliers. The means and standard deviations of the populations are, of course, assumed unknown.

(d) This is not an experiment. The groups studied were simply samples from the populations of infants whose mothers breast-fed them and of infants whose mothers formula-fed them. Treatments (breast-feeding versus formula-feeding) were not assigned to infants by the investigators.

Since this was not an experiment, we are much less certain that the observed results are due to effects (or lack of effects) of the treatments (breast-feeding versus formula-feeding). We must ask if any confounding factors exist. For example, mothers choose whether they breast-feed or formula-feed an infant. If this choice is in any way related to hemoglobin levels in infants (for example, suppose infants with low hemoglobin levels look less healthy than infants with high hemoglobin levels, and mothers decide to breast-feed versus formula-feed partly on the basis of how healthy the infant looks), than it would be confounded with the effects of the treatment. Such confounding effects might offset actual differences in breast-feeding from formula-feeding, leading to the non-significant results in the study.

Exercise 6.45

(a) The data can be presented graphically by means of stemplots, back-to-back stemplots, or histograms. We present the data for each group in separate histograms displayed below.

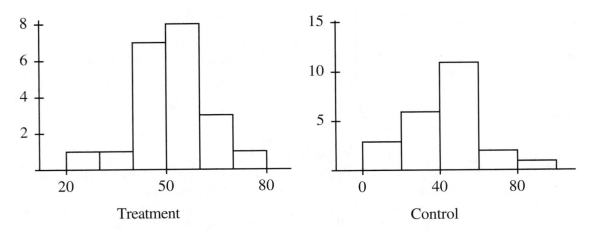

Neither histogram indicates there are any strong outliers and neither shows any strong skewness. Since the total sample size is 44, which is reasonably large, it is safe to use t procedures.

(b) If we calculate the mean and standard deviation for both groups, we obtain the following results

Group	n	$\bar{x}$	s
Treatment	21	51.48	11.01
Control	23	41.52	17.15

To carry out a test, we proceed as follows.

Step 1. Hypotheses. This is a two-sample problem. Let the treatment group be group 1 and the control group be group 2. Since we are interested in whether the data provide evidence that the treatment improves scores, we wish to test the hypotheses

$$H_0: \mu_1 = \mu_2$$
$$H_a: \mu_1 > \mu_2$$

Step 2. Test statistic. From the basic statistics given, we compute

$$t = \frac{\bar{x}_1 - \bar{x}_2}{\sqrt{\dfrac{s_1^2}{n_1} + \dfrac{s_2^2}{n_2}}} = \frac{51.48 - 41.52}{\sqrt{\dfrac{(11.01)^2}{21} + \dfrac{(17.15)^2}{23}}} = \frac{9.96}{\sqrt{18.56}} = 2.31$$

Step 3. P-value. The P-value for $t = 2.31$ is the area to the right of 2.31 under the t distribution curve with the minimum of $n_1 - 1 = 20$ and $n_2 - 1 = 22$ degrees of freedom. This minimum is 20. Using Table C, we search the $df = 20$ row for entries that bracket 2.31. We find

$$df = 20$$

p	.02	.01
t^*	2.197	2.528

The P-value is therefore between .02 and .01 for this one–sided test. The data provide strong evidence that the population represented by the treatment group has a higher mean DRP score than the population represented by the control group.

(c) Proper randomization is missing and is probably not possible (classes are probably fixed and students can't be rearranged). An entire class is assigned to a treatment, not individual students by random assignment. Any differences in the classes will be confounded with the effects of the treatment.

SECTION 6.3

(optional)

SECTION OVERVIEW

This section covers a basic inference method to use for the population standard deviation. The **F test** is not very robust. The F statistic has the **F distribution** which is noted as $F(j,k)$ where j and k are the degrees of freedom. Use this statistic only if you are sure the samples are SRSs and the populations are normal.

KEY CONCEPTS

The F statistic and its distribution

The F statistic is the ratio of the sample variances: $\frac{s_1^2}{s_2^2}$. The statistic has a distribution called the F distribution that actually has two kinds of degrees of freedom, one for each variance in the ratio, $n_1 - 1$ and $n_2 - 1$. There is a table, Table D, of F distributions we can use to find the critical values and cutoff points. Values close to 1 say that the variances are very similar so values far from 1 say that the variances are not the same.

The F test

The F test is used to test the hypotheses H_0: $\sigma_1 = \sigma_2$ vs. H_a: $\sigma_1 \neq \sigma_2$.

• Set up the statistic so that the larger s^2 is on the top.

• Compare the F value with the critical values from Table D.

• Double the significance level because this is a two-sided test.

SOLUTIONS TO SELECTED TEXT EXERCISES

Exercise 6.53

From Example 6.11 (DDT = group 1, Control = group 2), we know

$$s_1^2 = (6.34)^2 = 40.20$$
$$s_2^2 = (1.95)^2 = 3.80$$

To test

$$H_0: \sigma_1^2 = \sigma_2^2$$
$$H_a: \sigma_1^2 \neq \sigma_2^2$$

we compute the F statistic

$$F = \frac{s_1^2}{s_2^2} = \frac{40.20}{3.80} = 10.58$$

When H_0 is true F has the F distribution with $n_1 - 1 = 5$ and $n_2 - 1 = 5$ degrees of freedom. From Table D we find the .05 and .01 critical values are 5.05 and 10.97, respectively. Thus we would reject H_0 at the 5% level but not at the 1% level.

SOLUTIONS TO SELECTED TEXT REVIEW EXERCISES

Exercise 6.59

(a) The right test is the two-sample test. The two groups, the control group and the undermine group, consist of 45 women each. However, the 45 women in the two groups are different and there is no indication that the women in the two groups were matched in any way. Thus there is no natural way to pair observations from the two groups together. Such pairing is necessary in a matched pairs t test which uses the differences between matched pairs of observations to compute the t statistic.

(b) Both groups have 45 observations. If we use the conservative rule (degrees of freedom are one less than the smaller sample size), we would use $45 - 1 = 44$ degrees of freedom.

(c) Each group consists of 45 observations. These are large samples (both are > 40) and so it is safe to use the t procedures even thought the 7 point scale could not have a normal distribution. Note that with a 7 point scale (and sample means of around 4 or 5), the distribution of the observations could not have any extreme outliers and could not be strongly skewed, further reasons why the two sample t procedures should be reasonably accurate.

Exercise 6.63

To examine the distribution of the data we make separate histograms for the standard medium (control group) data and the nitrite medium data.

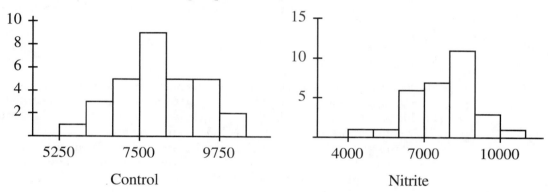

The distribution for the data from the standard medium is reasonably symmetric with no outliers. The distribution for the data from the nitrite medium is slightly skewed to the left, but there are no outliers.

To carry out a test, we proceed as follows.

Step 1. Hypotheses. This is a two-sample problem. Let the standard medium (control group) be group 1 and the nitrite medium be group 2. Let μ_1 and μ_2 be the means of the populations represented by groups 1 and 2, respectively. Since we are interested in whether the data provide evidence that nitrites decrease amino acid uptake, we wish to test the hypotheses

$$H_0: \mu_1 = \mu_2$$
$$H_a: \mu_1 > \mu_2$$

Step 2. Test statistic. From the data given, we compute the following values for the basic statistics

Group	n	$\bar{x}$	s
Standard medium	30	8072.93	1068.32
Nitrite medium	30	7807.30	1235.20

Thus the value of the test statistic t is

$$t = \frac{\bar{x}_1 - \bar{x}_2}{\sqrt{\dfrac{s_1^2}{n_1} + \dfrac{s_2^2}{n_2}}} = \frac{8072.93 - 7807.30}{\sqrt{\dfrac{(1068.32)^2}{30} + \dfrac{(1235.20)^2}{30}}} = \frac{265.63}{\sqrt{88900.89}} = 0.89$$

Step 3. P-value. The P-value for $t = 0.89$ is the area to the right of 0.89 under the t distribution curve with the minimum of $n_1 - 1 = 29$ and $n_2 - 1 = 29$ degrees of freedom. This minimum is 29. Using Table C, we search the $df = 29$ row for entries that bracket 0.89. We find

$$df = 29$$

p	.20	.15
t^*	0.854	1.055

The P-value is therefore between .20 and .15 for this one-sided test. The data do not provide strong evidence that the population represented by the nitrite group has a lower mean amino acid uptake than the population represented by the control group.

Exercise 6.67

It is not proper, nor necessary, to apply the one-sample t method to these data to give a 95% (or any level, for that matter) confidence interval for the mean population of an Indiana county. The data are, in fact, a census of all the counties in Indiana. Therefore the mean of the data is the mean population of an Indiana county. We know the true mean with 100% certainty, assuming the census data are accurate and we have made know arithmetic mistakes. There is no uncertainty in this problem. The one-sample t method would be appropriate if we had a small random sample of all the counties in Indiana.

Note that if the census data are inaccurate, the one-sample t method is still not appropriate. While we would be uncertain about the true value of the mean population of Indiana counties, our uncertainty would be due to non random errors arising from the way the data were collected, not because we only had a random sample from the population of all Indiana counties.

Exercise 6.69

(a) To carry out a test, we proceed as follows.

Step 1. Hypotheses. This is a two-sample problem. Let the pet dogs be group 1 and the clinic dogs be group 2. Let μ_1 and μ_2 be the means of the populations represented by groups 1 and 2, respectively. Since we are interested in whether the data provide evidence that pets (group 1) have a higher mean cholesterol level than clinic dogs (group 2), we wish to test the hypotheses

$$H_0: \mu_1 = \mu_2$$
$$H_a: \mu_1 > \mu_2$$

Step 2. Test statistic. From the data given, we compute the following values for the test statistic t

$$t = \frac{\bar{x}_1 - \bar{x}_2}{\sqrt{\dfrac{s_1^2}{n_1} + \dfrac{s_2^2}{n_2}}} = \frac{193 - 174}{\sqrt{\dfrac{(68)^2}{26} + \dfrac{(44)^2}{23}}} = \frac{19}{\sqrt{262.02}} = 1.17$$

Step 3. P-value. The P-value for $t = 1.17$ is the area to the right of 1.17 under the t distribution curve with the minimum of $n_1 - 1 = 25$ and $n_2 - 1 = 22$ degrees of freedom. This minimum is 22. Using Table C, we search the $df = 22$ row for entries that bracket 1.17. We find

$$df = 22$$

p	.15	.10
t^*	1.061	1.321

The P-value is therefore between .15 and .10 for this one-sided test. The data do not provide strong evidence that the population represented by the pet dogs has a higher mean cholesterol level than the population represented by the clinic dogs.

Exercise 6.73

To describe the distribution of city pollution levels, we look at a histogram of the data for city particulate levels from Table 6.2 (you could also look at a stemplot of the data).

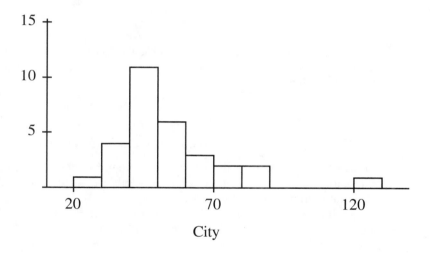

City

There is one outlier (123 grams on the 15th date). The distribution of the data, ignoring the outlier, is skewed to the right, but not unusually so.

The outlier suggests that the mean is not the best measure of center for these data. It also means that t-procedures are not appropriate, although the sample size of 30 is moderately large. Perhaps the best thing to do is to use the median as a measure of center of these data and give the 5 number summary along with a boxplot of the data. These results are presented below. The interrquartile range can be used as an estimate of the spread or variability in the data.

> Minimum = 23
> 1st quartile = 42
> Median = 48.5
> 3rd quartile = 62
> Maximum = 123
> Interquartile range = 20

A boxplot of the data is

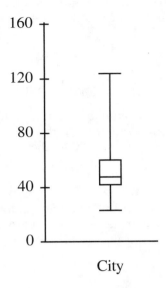

City

If we had gone ahead and calculated the mean of the data, the standard deviation of the data, and a 95% confidence interval for the mean, we would have obtained the following results:

Mean = 54.067

Standard deviation = 19.407

95% confidence interval for the mean = (46.82, 61.31)

Notice that the mean is larger than the mean (as one would expect with a large outlier) and the 95% confidence interval, while including the median of 48.5, tends to overstate where the center of the data is.

One might be tempted to remove the outlier from the data and compute the mean, along with a 95% confidence interval, for the remainder of the data. While this removes the effect of the outlier, it is hard to justify its removal. It appears to be a legitimate measurement and therefore should probably be retained. Use of the median, the five-number summary, and the interquartile range seems a better way to describe the data and summarize it.

Exercise 6.75

To examine the relationship between rural particulate level (the explanatory variable since it will be used to predict the city level) and city level (the response variable) we make a scatterplot of the data.

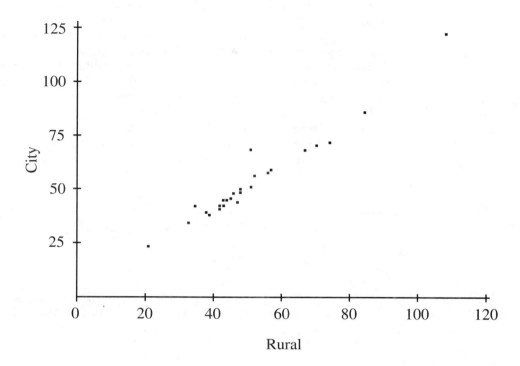

The scatterplot suggests a strong linear relation between rural and city levels. It appears that using the least-squares regression line for prediction will give approximately correct results over the range of the data. Even though there is an outlier in the upper right corner of the plot, it does appear to fall along the trend defined by the remaining points in the plot. It is probably safe to include it in subsequent analysis.

If we calculate the least-squares regression line we get the equation

city level = − 2.5801 + 1.0935(rural level)

and find that the value of r^2 is 0.951. Thus 95.1% of the observed variation in city pollution levels is accounted for by the straight-line relationship between city and rural pollution levels.

For a rural level reading of 88 (on the 14th date, and that is in the range of the data) we would predict a city level reading of

city level = − 2.5801 + 1.0935(88) = 93.6479

or approximately 94 grams(rounding off to the nearest gram since the data are measured to the nearest gram).

CASE STUDY

Not too long ago, it was reported that listening to Mozart improved performance on tests, at least temporarily. Perhaps other pleasant sensory inputs also improve mental abilities. For example, can pleasant aromas help a person learn better? Two researchers (Hirsch, A.R. and Johnston, L. H.) from the Smell & Taste Treatment and Research Foundation, Chicago, investigated whether the presence of a floral scent could improve a person's learning ability in certain situations. They found that most of their subjects showed greater improvement on their times through a pencil and paper maze when they were smelling a floral scent than when they weren't. In keeping with our principle of maintaining a healthy skepticism about such results until we have a chance to look at the data ourselves, we give information about how the study was conducted and provide the data below.

In the study, twenty-two people worked through a set of two pencil and paper mazes. The maze below is an example of one of them.

Maze

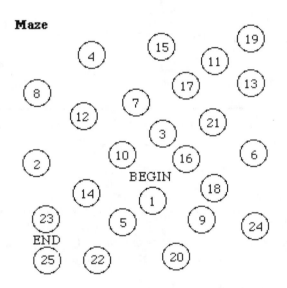

The maze is completed by connecting the 25 numbered circles in order.

Each subject worked both mazes six times, three times while wearing a floral-scented mask and three times wearing an unscented mask. The three trials for each mask closely followed one another. Individuals were randomly assigned to wear the floral mask on either their first three tries or their last three tries. Participants put on their masks one minute before starting the first trial in each group to minimize any distracting effect. Subjects recorded their opinion as to whether they found the scent inherently positive, inherently negative, or if they were indifferent to it. Testers measured the total length of time it took subjects to complete both mazes in each of the six trials. To ensure that the subjects could actually smell the floral scent, they were each given two tests measuring olfactory ability. The scores of one person were removed from the data set because he failed both of these smell tests.

The data for the remaining 21 subjects are given below.

	Opinion	Unscented trial 1	Unscented trial 2	Unscented trial 3	Scented trial 1	Scented trial 2	Scented trial 3
1.	Pos	38.4	27.7	25.7	53.1	30.6	30.2
2.	Neg	46.2	57.2	41.9	54.7	43.3	56.7
3.	Pos	72.5	57.9	51.9	74.2	53.4	42.4
4.	Neg	38.0	38.0	32.2	49.6	37.4	34.4
5.	Neg	82.8	57.9	64.7	53.6	48.6	44.8
6.	Pos	33.9	32.0	31.4	51.3	35.5	42.9
7.	Pos	50.4	40.6	40.1	44.1	46.9	42.7
8.	Pos	35.0	33.1	43.2	34.0	26.4	24.8
9.	Pos	32.8	26.8	33.9	34.5	25.1	25.1
10.	Indiff	60.1	53.2	40.4	59.1	87.1	59.2
11.	Pos	75.1	63.1	58.0	67.3	43.8	42.2
12.	Indiff	57.6	57.7	61.5	75.5	126.6	48.4
13.	Pos	55.5	63.3	44.6	41.1	41.8	32.0
14.	Indiff	49.5	45.8	35.3	52.2	53.8	48.1
15.	Indiff	40.9	35.7	37.2	28.3	26.0	33.7
16.	Pos	44.3	46.8	39.4	74.9	45.3	42.6
17.	Neg	93.8	91.9	77.4	77.5	55.8	54.9
18.	Neg	47.9	59.9	52.8	50.9	58.6	64.5
19.	Pos	75.2	54.1	63.6	70.1	44.0	43.1
20.	Neg	46.2	39.3	56.6	60.3	47.8	52.8
21.	Neg	56.3	45.8	58.9	59.9	36.8	44.3

We notice several things about this study. First, it was a randomized, comparative experiment. Subjects received both treatments (floral versus unscented masks) and were assigned which treatment they received first by randomization. Inferences were made by comparing the effects of the two treatments. Second, this was a matched pairs study with the results under the two treatments for each subject serving as the matched pair. Third, the researchers tried to control for confounding factors by making experimental conditions similar in both trials. Subjects wore masks and did similar tasks under similar conditions under both treatments. Also, subjects were tested to make sure they could actually smell the floral scent. They were also asked whether they liked the scent. This information could be used to determine if just the presence of the floral scent (regardless of whether the subject finds it pleasant) has an effect, or if an effect occurs only for subjects who find the scent pleasant.

Let us now think about what question the researchers wished to answer. The stated goal was to see if the presence of a floral scent could improve a person's learning ability in certain situations (in this case, working through a maze). Comparing the mean time to complete the mazes while wearing the scented mask with the mean time while wearing the unscented mask might tell us if the presence of the floral scent improves mean performance in completing the maze, but would not give us information about improvement in the ability to learn the maze. How might the data be used to measure "learning"? Improvement in the time to complete the two mazes over the course of the three trials made while wearing a particular mask might indicate that learning has taken place. Thus we could compute the difference between the times to complete the first and third trial and take this to be a measure of the amount of learning that has taken place. For a particular treatment (either wearing the scented mask or the unscented mask) we would define

amount of learning = time taken in first trial – time taken in third trial

Better yet, we might divide this difference by the time taken in the first trial to get the percent or relative improvement, i. e.,

$$\text{relative amount of learning } = \frac{\text{time taken in first trial - time taken in third trial}}{\text{time taken in first trial}}$$

Using the relative improvement puts all results on a common (unitless) scale and makes for fairer comparisons. We could compare the relative improvement while wearing the scented mask with the relative improvement while wearing the unscented mask to see if there is any difference in the relative amount of learning that takes place under these two treatments.

If we use relative amount of learning as our measure and compute this for each subject and for both treatments, our data becomes the following.

	Opinion	Relative improvement (Unscented)	Relative improvement (Scented)	Difference (Unscented – scented)
1.	Pos	0.33072917	0.43126177	–0.10053260
2.	Neg	0.09307359	–0.03656307	0.12963666
3.	Pos	0.28413793	0.42857143	–0.14443350
4.	Neg	0.15263158	0.30645161	–0.15382003
5.	Neg	0.21859903	0.16417910	0.05441993
6.	Pos	0.07374631	0.16374269	–0.08999638
7.	Pos	0.20436508	0.03174603	0.17261905
8.	Pos	–0.23428571	0.27058824	–0.50487395
9.	Pos	–0.03353659	0.27246377	–0.30600035
10.	Indiff	0.32778702	–0.00169205	0.32947907
11.	Pos	0.22769640	0.37295691	–0.14526050
12.	Indiff	–0.06770833	0.35894040	–0.42664873
13.	Pos	0.19639640	0.22141119	–0.02501480
14.	Indiff	0.28686869	0.07854406	0.20832463
15.	Indiff	0.09046455	–0.19081272	0.28127727
16.	Pos	0.11060948	0.43124166	–0.32063217
17.	Neg	0.17484009	0.29161290	–0.11677282
18.	Neg	–0.10229645	–0.26719057	0.16489412
19.	Pos	0.15425532	0.38516405	–0.23090873
20.	Neg	–0.22510823	0.12437811	–0.34948633
21.	Neg	–0.04618117	0.26043406	–0.30661523

We might consider using a matched pairs t–test to see if the presence of the floral scent improves learning. A histogram of the differences (last column of the data above) is

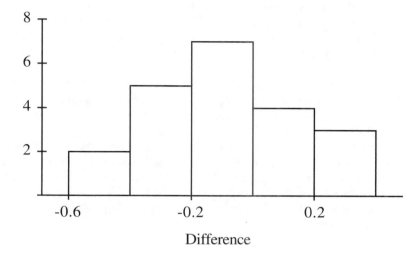

Difference

This shows that there are no outliers or substantial skewness in the data (in fact, it looks reasonably normal), so the t-test appears safe to use.

Let μ be the mean difference in relative improvement in learning the maze between the unscented and scented conditions for the population represented by the subjects in the study. The scented condition had a higher mean relative improvement in learning, as compared with the unscented condition, if $\mu < 0$. Thus to test the researcher's hypothesis, we might use the matched pairs t-test to test the hypotheses

$$H_0: \mu = 0$$
$$H_a: \mu < 0$$

If one computes the summary statistics for the differences, one gets

$$n = 21, \ \bar{x} = -0.08954, s = 0.23879$$

resulting in a t of

$$t = \frac{\bar{x} - \mu_0}{s/\sqrt{n}} = \frac{-0.08954 - 0}{0.23879 / \sqrt{21}} = 1.718$$

This has a P-value of between 0.05 and 0.10 (you should check this using Table C in the text). In fact, statistical software gives a P-value of 0.0506. This is moderate but not strong evidence against the null hypothesis. In other words, the data give some evidence that the relative improvement in learning the maze is greater while wearing the floral scented mask than while wearing the unscented mask.

Actually, it might make more sense to test the hypothesis only using the data for the subjects that found the floral scent positive. Remember, the hypothesis is that learning will be improved by the presence of a pleasant scent. Only data for those subjects that liked the floral scent actually give us information about this hypothesis. For those who liked the floral scent, the summary statistics for the differences are

$$n = 10, \ \bar{x} = -0.16950, s = 0.18497$$

resulting in a t of

$$t = \frac{\bar{x} - \mu_0}{s/\sqrt{n}} = \frac{-0.16950 - 0}{0.18497 / \sqrt{10}} = 2.90$$

This has a *P*-value between 0.010 and 0.005 (once again, you should check this using Table C in the text). This is reasonably strong evidence that greater relative improvement in learning the maze occurred in the presence of the pleasant floral scent than in the unscented condition.

What can we conclude from this study? While the results were "statistically significant" (at least for those who found the scent pleasant), several things should be kept in mind. First, it is not clear what population of individuals the subjects represent. We are not told if the subjects were a sample from some larger population, so we need to be cautious about how far the results of the study can be generalized. Second, this experiment is artificial, as experiments on human subjects often tend to be. The experiment only demonstrated improvement in the ability to work a pencil and paper maze in a controlled setting while wearing a scented mask versus wearing an unscented mask. This does not necessarily mean that the presence of a pleasant scent would improve one's ability to learn other things. Nevertheless, the results of the study are intriguing. While spraying your room with a pleasant smelling air freshener is probably not a satisfactory replacement for serious study before an exam, it probably won't hurt you either. And maybe it will even enhance the time you do spend studying!

CHAPTER 7

INFERENCE FOR PROPORTIONS

CHAPTER OVERVIEW

In each situation there are many questions that can be asked. Every question may require a different approach. Here are three questions you could encounter: Is the mean number of words in a sentence written by a person with a high school diploma different from the mean number of words in a sentence written by a person with a college degree? Does a new nail manufacturing process have less variation in nail lengths than the old process? What percentage of speeding drivers slow down after learning there is a radar gun nearby? So far we can answer the first two questions using the tools we acquired in Chapter 6. The third type of question is the subject of Chapter 7: inference about population proportions (or percentages).

The goal of this chapter is to help us recognize when to use the techniques for population proportion and how to use the techniques properly. We shall see the similarities between these new methods and the ones we have already learned. Here in chapter 7 we will use a z statistic again but in a new setting. The chapter covers inference for a single population and for comparing two populations. (Tools for more than two populations are saved until Chapter 8.? Remember, anyone can plug numbers into a formula and get an answer, but it takes an educated brain to make sense of the answer and use the information well. All the tools we will learn to use come with warning labels, so pay attention to them. As before, always check to be sure that the assumptions about the data are met before carrying out any type of inference.

SECTION 7.1

SECTION OVERVIEW

In the rest of the chapter we will be talking about the number of "successes," consider a **success** to be the outcome you are interested in. We will need to count the number of success out of the number of observations, n. The statistic that we will use to measure the population proportion is the **sample proportion,** $\hat{p}$. Chapter 4, section 3, introduced us to this statistic. This time we won't know the population proportion and so we will use the **standard error** to estimate the standard deviation. All the methods in this section assume the population is large and so we will use the normal approximation methods. Here we will also see how using planned inference can guide the design of the experiment or study.

KEY CONCEPTS

Situation assumptions

1. The data are an SRS. This is very important if we want to extend our conclusions to the population.

2. The population is at least 10 times larger than the sample size, n.

3. For a significance test? when testing H_0: $p = p_0$, np_0 and $n(1 - p_0)$ must both be larger than 10.

4. For a confidence interval: $n\hat{p}$, the number of success, and $n(1 - \hat{p})$, the number of failures, must both be larger than 10.

Details on the sampling distribution of $\hat{p}$

$$\hat{p} = \frac{number\ of\ success\ in\ the\ sample}{number\ of\ observations}$$

- The shape of the distribution is close to normal and gets more and more normal as the sample size, n, increases.
- The mean of the distribution is the population proportion, p.
- The standard deviation gets smaller as the sample gets larger.
- The standard deviation can be calculated using the following formula: $\sqrt{\dfrac{p(1-p)}{n}}$, where n is the sample size.
- Since the population proportion will be unknown, we must estimate the standard-deviation by the standard error. When testing the hypothesis H_0: $p = p_0$, replace p with p_0. When creating a confidence interval, use $\hat{p}$, the sample proportion, in place of p.

Confidence intervals using $\hat{p}$

The approximate level C confidence interval for the population proportion, p, is
$\hat{p} \pm z^* \sqrt{\dfrac{\hat{p}(1-\hat{p})}{n}}$, where z^* is the upper $(1-C)/2$ critical value from the standard normal table.

Hypothesis tests using $\hat{p}$

1. Write out the alternative hypotheses you want to test. $H_0: p = p_0$ is the null hypothesis. The alternative can be one of the following:

$$H_a: p > p_0$$
$$H_a: p < p_0$$
$$H_a: p \neq p_0$$

2. Calculate $\hat{p}$ and then find z. $z = \dfrac{\hat{p} - p_0}{\sqrt{\dfrac{p_0(1-p_0)}{n}}}$. This is the standardized value of $\hat{p}$.

3. Find the P-value that is appropriate depending on how you wrote the alternative hypothesis. As with any probability calculation, you may find it helpful to draw a picture of the area you are looking for.

If H_a? $p > p_0$ then find $P(Z \geq z)$.
If H_a? $p < p_0$ then find $P(Z \leq z)$.
If H_a? $p \neq p_0$ then find $2P(Z \geq |z|)$.

Z has a standard normal distribution.

4. State your conclusion in terms of the question. Don't just give a P-value.

Sample size and $\hat{p}$

Confidence intervals all have the form estimate $\pm$ margin of error. The margin of error consists of a critical value and the standard error or standard deviation. If the margin of error involves n we can turn things around and use a fixed margin of error to tell us the sample size needed to achieve it. For the sample proportion, $n = \left(\dfrac{z^*}{m}\right)^2 p^*(1-p^*)$, where m is the fixed value you have chosen for the margin of error. If you have a good guess as to the value of p^* then use 0.5, this will give the maximum sample size.

SOLUTIONS TO SELECTED TEXT EXERCISES

Exercise 7.3

(a) The population is presumably all living alumni of the college. The parameter p is the proportion of these alumni who support firing the coach.

(b) The statistic $\hat{p}$ is the proportion in the SRS that support firing the coach. It has value

$$\hat{p} = 76/200 = 0.38.$$

Exercise 7.5

(a) • The 10 tosses might be regarded as an SRS from the population of all possible tosses of the coin, assuming nothing unusual was done while tossing the coin.

   • This "hypothetical" population is infinite, so the population size is certainly at least ten times the size of the sample of 10 tosses.

   • For testing the hypothesis H_0: $p = 0.5$, we require $n(0.5)$ and $n(1 - 0.5)$ to both be 10 or more. Here $n = 10$, so $n(0.5) = 10(0.5) = 5$ and $n(1 - 0.5) = 10(1 - 0.5) = 5$, both of which are *smaller* than 10.

We conclude that it is *not* safe to use the methods described in the section.

(b) • The sample is an SRS from the population of all living alumni.

   • The population size is 15,000 which is certainly at least ten times the size of the sample of 200 alumni.

   • For testing the hypothesis H_0: $p = 0.99$, we require $n(0.99)$ and $n(1 - 0.99)$ to both be 10 or more. Here $n = 200$, so $n(0.99) = 200(0.99) = 198$ but $n(1 - 0.99) = 200(1 - 0.99) = 2$. The latter is smaller than 10.

We conclude that it is *not* safe to use the methods described in the section.

(c) • The sample is an SRS from the population of the 250 students in the course.

   • The population size is 250 which is certainly at least ten times the size of the sample of 20 students.

   • For testing the hypothesis H_0: $p = 0.5$, we require $n(0.5)$ and $n(1 - 0.5)$ to both be 10 or more. Here $n = 20$, so $n(0.5) = 20(0.5) = 10$ and $n(1 - 0.5) = 20(1 - 0.5) = 10$. Both are at least 10.

We conclude that it is safe to use the methods described in the section.

Exercise 7.9

(a) *Step 1: Hypotheses.* The null hypothesis is that the probability p of coming up heads is 0.5 versus the two-sided alternative. Thus we test the hypotheses

$$H_0: p = 0.5$$
$$H_a: p \neq 0.5$$

The null hypothesis gives the value of $p_0 = 0.5$. If we regard Karl Pearson's $n = 24,000$ tosses as an SRS from an infinite population, the population size is certainly at least ten times the sample size. We also note that $np_0 = 24,000(0.5) = 12,000$ and $n(1 - p_0) = 24,000(1 - 0.5) = 12,000$, so it is safe to use the procedures of this section.

Step 2. Test statistic. We notice that $\hat{p} = 12,012/24,000 = 0.5005$. The z test statistic is therefore

$$z = \frac{\hat{p} - p_0}{\sqrt{\frac{p_0(1 - p_0)}{n}}} = \frac{0.5005 - 0.5}{\sqrt{\frac{0.5(1 - 0.5)}{24,000}}} = \frac{0.005}{0.0032275} = 1.55$$

Step 3. P-value. Because the test is two-sided, the *P*-value is the area under a standard normal curve more than 1.55 away from 0 in either direction. The figure below shows this area.

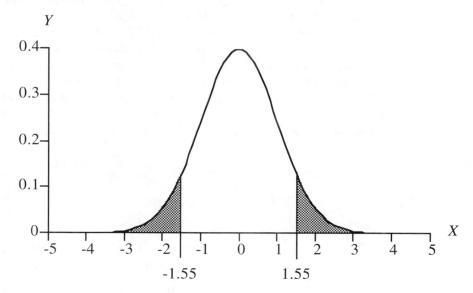

From Table A we find that the area below −0.16 is 0.4364. The *P*-value is twice this area.

$$P = 2(0.4364) = 0.8728$$

Conclusion. A proportion of heads as far from 1/2 as obtained by Pearson would happen 87.28% of the time when a balanced coin is tossed 24,000 times. Pearson's results are not strong evidence that the coin is unbalanced.

(b) A 99% confidence interval for the probability *p* of heads for Pearson's coin uses the standard normal critical value $z^* = 2.576$ from Table C (look in the bottom row for standard normal critical values). The confidence interval is

$$\hat{p} \pm z^* \sqrt{\frac{\hat{p}(1-\hat{p})}{n}} = 0.5005 \pm 2.576 \sqrt{\frac{0.5005(1-0.5005)}{24,000}}$$

$$= 0.5005 \pm 0.0083$$
$$= (0.4922, 0.5088)$$

Exercise 7.11

We start with the guess that $p^* = 0.75$. For 95% confidence we use $z^* = 1.96$. The sample size we need for a margin of error $m = 0.04$ is thus

$$n = \left(\frac{z^*}{m}\right)^2 p^*(1-p^*) = \left(\frac{1.96}{0.04}\right)^2 0.75(1-0.75) = 450.1875$$

We round this up to $n = 451$. Thus a sample of size 451 is needed to estimate the proportion of Americans with at least one Italian grandparent who can taste PTC to within ± .04 with 95% confidence.

Exercise 7.15

(a) We assume that the population of all food and drink businesses in the 12 counties in central Indiana (which presumably includes Indianapolis) is at least ten times the size of the sample of $n = 148$ businesses. We notice that $\hat{p} = 22/148 = 0.1486$ so that $n\hat{p} = 148(0.1486) = 22$ and $n(1-\hat{p}) = 148(1 - 0.1486) = 126$. If we pretend that the $n = 148$ businesses that did respond are an SRS from the population of all food and drink businesses in the 12 counties in central Indiana (an assumption that is not valid, as we shall see in (c)), then we may safely apply the methods of this chapter. A 95% confidence interval for the probability p of heads for Pearson's coin uses the standard normal critical value $z^* = 1.96$ from Table C (look in the bottom row for standard normal critical values). The confidence interval is

$$\hat{p} \pm z^* \sqrt{\frac{\hat{p}(1-\hat{p})}{n}} = 0.1486 \pm 1.96 \sqrt{\frac{0.1486(1-0.1486)}{148}}$$

$$= 0.1486 \pm 0.0573$$

$$= (0.0913 , 0.2059)$$

(b) We start with the guess that $p^* = 0.1486$. For 95% confidence we use $z^* = 1.96$. The sample size we need for a margin of error $m = 0.04$ is thus

$$n = \left(\frac{z^*}{m}\right)^2 p^*(1 - p^*) = \left(\frac{1.96}{0.04}\right)^2 0.1486(1 - 0.1486) = 303.77.$$

We round this up to $n = 304$. Thus a sample of size 304 is needed to reduce the margin of error to .04 with 95% confidence.

(c) There are several things that make it unlikely that their findings describe the population of all small businesses. First, they only sampled small food and drink businesses. It is not clear that these are representative of all small businesses. Second, they only sampled from the Yellow Pages for a twelve-county area in central Indiana. It is not clear to what extent this region is representative of a larger geographic area. Finally, there was a large percentage (45%) of nonresponses. We do not know in what way (if any) these nonrespondents differed from the businesses actually contacted. One possibility is that nonrespondents were businesses that were failing or doing poorly and hence were not open. If this is the case, than the authors' findings are likely to underestimate the probability of failing in three years. In addition, the large percentage of nonrespondents means that the sample obtained by the authors is not an SRS and hence that the statistical methods used in part (a) are not safe to use. We cannot trust the confidence interval calculated.

At best, the study may describe the probability of failure in three years of the population of food and drink businesses that are currently healthy and located in areas similar to the twelve-county region sampled in central Indiana.

SECTION 7.2

SECTION OVERVIEW

When the situation involves comparing two populations or the responses to two treatments based on two independent samples we need to use tools that fit the situation. Just as in Chapter 6 we compared two populations using the difference of the sample statistics, we will use the difference of the sample proportions, $\hat{p}_1 - \hat{p}_2$, as our statistic. As in Chapter 6, the subscript on the variable will denote which population it is from. You can pick either population to be 1.

KEY CONCEPTS

Sampling distribution of $\hat{p}_1 - \hat{p}_2$

- The mean of the distribution is $p_1 - p_2$.
- The variance is $\dfrac{p_1(1-p_1)}{n_1} + \dfrac{p_2(1-p_2)}{n_2}$.
- The distribution is approximately normal if the sample sizes are large.

- For the confidence intervals and significance tests we will use standard errors to estimate the standard deviation.

- Use these methods only if the population is 10 times larger than the sample and $n_1\hat{p}_1$, $n_1(1-\hat{p}_1)$, $n_2\hat{p}_2$, and $n_2(1-\hat{p}_2)$ are all bigger than 5.

Confidence intervals using $\hat{p}_1 - \hat{p}_2$

For confidence intervals we need to use a standard error for the statistic. The standard error is $\sqrt{\dfrac{\hat{p}_1(1-\hat{p}_1)}{n_1} + \dfrac{\hat{p}_2(1-\hat{p}_2)}{n_2}}$, we just replaced the unknown population parameters with the sample statistics. The confidence interval has the usual form of estimate $\pm$ z*standard error of the estimate. So the approximate level C confidence interval is

$$(\hat{p}_1 - \hat{p}_2) \pm z^* \sqrt{\dfrac{\hat{p}_1(1-\hat{p}_1)}{n_1} + \dfrac{\hat{p}_2(1-\hat{p}_2)}{n_2}}$$

As always, z^* is the $(1 - C)/2$ standard normal critical value from the table.

Significance tests using $\hat{p}_1 - \hat{p}_2$

1. Write out the alternative hypotheses you want to test. $H_0: p_1 = p_2$ (or $H_0: p_1 - p_2 = 0$) is the null hypothesis. The alternative can be one of the following:

$$H_a? \; p_1 > p_2$$
$$H_a? \; p_1 < p_2$$
$$H_a? \; p_1 \neq p_2$$

2. Find z. $z = \dfrac{\hat{p}_1 - \hat{p}_2}{\sqrt{\hat{p}(1-\hat{p})(\dfrac{1}{n_1} + \dfrac{1}{n_2})}}$. This is the standardized value of $\hat{p}_1 - \hat{p}_2$.

The value $\hat{p}$ is the pooled sample proportion. Combine the samples and count the total number of successes out of the total number of observations.

3. Find the *P*-value that is appropriate depending on how you wrote the alternative hypothesis. As with any probability calculation, you may find it helpful to draw a picture of the area you are looking for. Use the standard normal table to get the values.

If H_a? $p_1 > p_2$ then find $P(Z \geq z)$.

If H_a? $p_1 < p_2$ then find $P(Z \leq z)$.

If H_a? $p_1 \neq p_2$ then find $2P(Z \geq |z|)$.

Z has a standard normal distribution.

4. State your conclusion in terms of the question. Don't just give a *P*-value?

SOLUTIONS TO SELECTED TEXT EXERCISES

Exercise 7.19

(a) Here, the sample sizes are for Detroit area white Protestants (population 1) n_1 = 267 and for Detroit area white Catholics (population 2) n_2 = 230. The sample proportions supporting the right to make speeches in favor of communism are $\hat{p}_1$ = 104/267 = 0.3895 and $\hat{p}_2$ = 75/230 = 0.3261. For a 95% confidence interval we use z^* = 1.96. The resulting interval is

$$(\hat{p}_1 - \hat{p}_2) \pm z^* \sqrt{\frac{\hat{p}_1(1-\hat{p}_1)}{n_1} + \frac{\hat{p}_2(1-\hat{p}_2)}{n_2}} =$$

$$= (0.3895 - 0.3261) \pm (1.96) \sqrt{\frac{0.3895(1-0.3895)}{267} + \frac{0.3261(1-0.3261)}{230}}$$

$$= 0.0634 \pm (1.96)(0.043)$$
$$= 0.0634 \pm 0.0843$$
$$= (-0.0209, 0.1477)$$

(b) We are told that the sample is basically a simple random sample of the population of the Detroit metropolitan area. Certainly the population is much larger than the sample sizes. We note that $n_1 \hat{p}_1$ = 267(0.3895) = 104, $n_1(1 - \hat{p}_1)$ = 267(1 − 0.3895) = 163, $n_2 \hat{p}_2$ = 230(0.3261) = 75, and $n_2(1 - \hat{p}_2)$ = 230(1 − 0.3261) = 155. All are at least 5, so it would appear safe to use the *z* confidence interval.

Exercise 7.21

(a) Let Detroit area white Protestants be population 1 and Detroit area white Catholics be population 2. Let p_1 and p_2 be the population proportions of Protestants and Catholics, respectively, who feel that the government was not doing enough in the areas stated. We wish to see if there is evidence that the white Protestants and white Catholics differed on the issue, i.e., were the proportions who feel that the government was not doing enough in the areas stated differently. The hypotheses we wish to test are therefore

$$H_0: p_1 = p_2$$
$$H_a: p_1 \neq p_2$$

The sample sizes are, for Detroit area white Protestants n_1 = 267 and for Detroit area white Catholics n_2 = 230. The sample proportions are $\hat{p}_1$ = 161/267 = 0.603 and

$\hat{p}_2 = 136/230 = 0.591$. We are told that the sample is basically a simple random sample of the population of the Detroit metropolitan area. Certainly the population is much larger than the sample sizes. We note that $n_1 \hat{p}_1 = 267(0.603) = 161$, $n_1(1 - \hat{p}_1) = 267(1 - 0.603) = 106$, $n_2 \hat{p}_2 = 230(0.591) = 136$, and $n_2(1 - \hat{p}_2) = 230(1 - 0.591) = 94$. All are at least 5, so it would appear safe to use the z test.

(b) The pooled proportion of white Protestants and Catholics who felt that the government was not doing enough in the areas stated was

$$\hat{p} = \frac{\text{count of all no answers for both Protestants and Catholics}}{\text{count of observations in both samples combined}}$$

$$= \frac{161 + 136}{267 + 230}$$

$$= \frac{297}{497} = 0.598$$

The z test statistic is

$$z = \frac{\hat{p}_1 - \hat{p}_2}{\sqrt{\hat{p}(1-\hat{p})\left(\dfrac{1}{n_1} + \dfrac{1}{n_2}\right)}}$$

$$= \frac{0.603 - 0.591}{\sqrt{0.598(1-0.598)\left(\dfrac{1}{267} + \dfrac{1}{230}\right)}}$$

$$= \frac{0.012}{0.044}$$

$$= 0.27$$

Because the test is two-sided, the P-value is the area under a standard normal curve more than 0.27 away from 0 in either direction. The figure below shows this area.

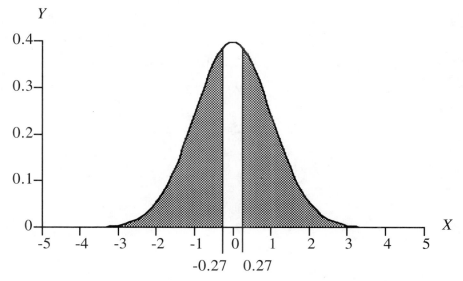

From Table A we find that the area below -0.27 is 0.3936. The P-value is twice this area.

$$P = 2(0.3936) = 0.7872$$

We conclude that the data does not provide strong evidence that white Protestants and white Catholics differed on the issue.

Exercise 7.23

(a) We want to test the claim that taking AZT lowers the proportion of infected people who will develop AIDS in a given time. Let those who received AZT be group 1 and who received the placebo be group 2. Let p_1 be the probability that a randomly selected subject in the study who is treated with AZT will develop AIDS and p_2 be the probability that a randomly selected subject in the study who is treated with a placebo will develop AIDS in a given time. The hypotheses we wish to test are therefore

$$H_1: p_1 = p_2$$
$$H_a: p_1 < p_2$$

The sample sizes are, for group 1, $n_1 = 435$ and for group 2, $n_2 = 435$. The sample proportions are $\hat{p}_1 = 17/435 = 0.039$ and $\hat{p}_2 = 38/435 = 0.087$. We are told that assignments were made at random. We note that $n_1\hat{p}_1 = 435(0.039) = 17$, $n_1(1 - \hat{p}_1) = 435(1 - 0.039) = 418$, $n_2\hat{p}_2 = 435(0.087) = 38$, and $n_2(1 - \hat{p}_2) = 435(1 - 0.087) = 397$. All are at least 5, so it would appear safe to use the z test.

(b) The pooled proportion of subjects who developed AIDS was

$$\hat{p} = \frac{\text{count of all subjects who developed AIDS}}{\text{count of observations in both groups combined}}$$

$$= \frac{17 + 38}{435 + 435}$$

$$= \frac{55}{870} = 0.063$$

The z test statistic is

$$z = \frac{\hat{p}_1 - \hat{p}_2}{\sqrt{\hat{p}(1-\hat{p})\left(\dfrac{1}{n_1} + \dfrac{1}{n_2}\right)}}$$

$$= \frac{0.039 - 0.087}{\sqrt{0.063(1-0.063)\left(\dfrac{1}{435} + \dfrac{1}{435}\right)}}$$

$$= \frac{-0.048}{0.016}$$

$$= -3.00$$

Because the test is one-sided, the P-value is the area under a standard normal curve below −3.00. The figure below shows this area.

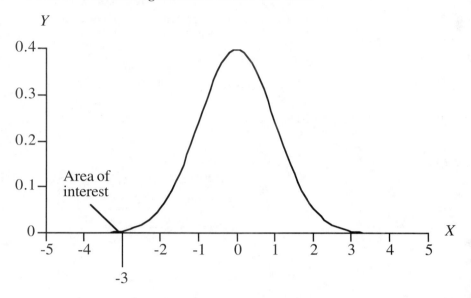

From Table A we find that the area below −3.00 is 0.0013. The P-value is thus

$$P = 0.0013$$

This is strong evidence that AZT lowers the proportion of infected people who will develop AIDS in a given period of time.

(c) Double blind means that neither the subjects in the study nor the physicians responsible for deciding whether a subject had developed AIDS knew which treatment the subjects received.

Exercise 7.29

Let the aspirin group be group 1 and the placebo group, group 2. Let p_1 and p_2 denote the true proportions for the outcome of interest for groups 1 and 2, respectively. For each outcome we test the two-sided hypotheses

$$H_0: p_1 = p_2$$
$$H_a: p_1 \neq p_2$$

The test for each outcome is calculated below. We omit a few details, such as diagrams, in the interest of space.

Fatal heart attacks

The sample sizes are, for group 1, $n_1 = 11,037$ and for group 2, $n_2 = 11,034$. The sample proportions are $\hat{p}_1 = 10/11,037 = 0.000906$ and $\hat{p}_2 = 26/11,034 = 0.002356$. We are told that assignments were made at random. We note that $n_1 \hat{p}_1 = 11,037(0.000906) = 10$, $n_1(1 - \hat{p}_1) = 11,037(1 - 0.000906) = 11,027$, $n_2 \hat{p}_2 = 11,034(0.002356) = 26$, and $n_2(1 - \hat{p}_2) = 11,034(1 - 0.002356) = 11,008$. All are at least 5, so it would appear safe to use the z test.

The pooled sample proportion is

$$\hat{p} = \frac{\text{count of all subjects who had fatal heart attacks}}{\text{count of observations in both groups combined}}$$

$$= \frac{10 + 26}{11{,}037 + 11{,}034}$$

$$= \frac{36}{22{,}071} = 0.001631$$

Thus the z test statistic is

$$z = \frac{\hat{p}_1 - \hat{p}_2}{\sqrt{\hat{p}(1-\hat{p})\left(\dfrac{1}{n_1} + \dfrac{1}{n_2}\right)}}$$

$$= \frac{0.000906 - 0.002356}{\sqrt{0.001631(1-0.001631)\left(\dfrac{1}{11{,}037} + \dfrac{1}{11{,}034}\right)}}$$

$$= \frac{-0.001450}{0.000543}$$

$$= -2.67$$

Because the test is two-sided, the P-value is the area under a standard normal curve more than 2.67 away from 0 in either direction; below −2.67 and above 2.67. From Table A we find that the area below −2.67 is 0.0038. The P-value is twice this area.

$$P = 2(0.0038) = 0.0076$$

We conclude that there is strong evidence that the aspirin and placebo groups have a different proportion of fatal heart attacks.

Non-fatal heart attacks

The sample sizes are, for group 1, $n_1 = 11{,}037$ and for group 2, $n_2 = 11{,}034$. The sample proportions are $\hat{p}_1 = 129/11{,}037 = 0.01169$ and $\hat{p}_2 = 213/11{,}034 = 0.01930$. We are told that assignments were made at random. We note that $n_1\hat{p}_1 = 11{,}037(0.01169) = 129$, $n_1(1 - \hat{p}_1) = 11{,}037(1 - 0.01169) = 10{,}908$, $n_2\hat{p}_2 = 11{,}034(0.01930) = 213$, and $n_2(1 - \hat{p}_2) = 11{,}034(1 - 0.01930) = 10{,}821$. All are at least 5, so it would appear safe to use the z test.

The pooled sample proportion is

$$\hat{p} = \frac{\text{count of all subjects who had non-fatal heart attacks}}{\text{count of observations in both groups combined}}$$

$$= \frac{129 + 213}{11{,}037 + 11{,}034}$$

$$= \frac{342}{22{,}071} = 0.01550$$

Thus the z test statistic is

$$z = \frac{\hat{p}_1 - \hat{p}_2}{\sqrt{\hat{p}(1-\hat{p})\left(\dfrac{1}{n_1} + \dfrac{1}{n_2}\right)}}$$

$$= \frac{0.01169 - 0.01930}{\sqrt{0.01550(1-0.01550)\left(\dfrac{1}{11,037} + \dfrac{1}{11,034}\right)}}$$

$$= \frac{-0.00761}{0.00166}$$

$$= -4.58$$

Because the test is two-sided, the P-value is the area under a standard normal curve more than 4.58 away from 0 in either direction; below −4.58 and above 4.58. From Table A we find that the area below −4.58 is off the table. The P-value turns out to be below .00001. We conclude that there is very strong evidence that the aspirin and placebo groups have a different proportion of non-fatal heart attacks.

Strokes

The sample sizes are, for group 1, n_1 = 11,037 and for group 2, n_2 = 11,034. The sample proportions are $\hat{p}_1$ = 119/11,037 = 0.01078 and $\hat{p}_2$ = 98/11,034 = 0.00888. We are told that assignments were made at random. We note that $n_1 \hat{p}_1$ = 11,037(0.01078) = 119, $n_1(1 - \hat{p}_1)$ = 11,037(1 − 0.01078) = 10,918, $n_2 \hat{p}_2$ = 11,034(0.00888) = 98, and $n_2(1 - \hat{p}_2)$ = 11,034(1 − 0.00888) = 10,936. All are at least 5, so it would appear safe to use the z test.

The pooled sample proportion is

$$\hat{p} = \frac{\text{count of all subjects who had strokes}}{\text{count of observations in both groups combined}}$$

$$= \frac{119 + 98}{11,037 + 11,034}$$

$$= \frac{217}{22,071} = 0.009832$$

Thus the z test statistic is

$$z = \frac{\hat{p}_1 - \hat{p}_2}{\sqrt{\hat{p}(1-\hat{p})\left(\dfrac{1}{n_1} + \dfrac{1}{n_2}\right)}}$$

$$= \frac{0.01078 - 0.00888}{\sqrt{0.009832(1-0.009832)\left(\dfrac{1}{11,037} + \dfrac{1}{11,034}\right)}}$$

$$= \frac{0.00190}{0.00133}$$

$$= 1.43$$

Because the test is two-sided, the P-value is the area under a standard normal curve more than 1.43 away from 0 in either direction below −1.43 and above 1.43. From Table A we find that the area below −1.43 is 0.0764. The P-value is twice this area.

$$P = 2(0.0764) = 0.1528$$

We conclude that there is not strong evidence that the aspirin and placebo groups have a different proportion of strokes.

Overall summary

There is strong evidence that there is a difference in the proportion of fatal and non-fatal heart attacks experienced by the aspirin and placebo groups. There is not strong evidence of a difference in the proportion of strokes experienced by the aspirin and placebo groups.

Note: When we make several comparisons, as we do here, we must be careful interpreting our results. We know, for example, that if we make 100 comparisons between groups which do not really differ, we would nevertheless expect to see about 5 comparisons appear significant at the 0.05 level. There are methods that adjust for the fact that we are making several comparisons. These are called multiple comparison procedures. If these are applied in this example, we would reach the same overall conclusions as we did with our three z-tests, but the *P*-values would not be quite as small. In general, if you are confronted with a problem requiring that you make many comparisons and reach some overall conclusion, consult a statistician.

SOLUTIONS TO SELECTED TEXT REVIEW EXERCISES

Exercise 7.33

In order for the procedures of this chapter to be safe, samples should be simple random samples from the populations of interest. Call-in polls are not simple random samples. You will recall from Chapter 3 that call-in surveys are often biased. Both the value of 81% and the calculations that produce the confidence interval should not be trusted. Fancy statistical procedures do not fix poorly designed surveys.

Exercise 7.37

(a)*Step 1: Hypotheses.* The null hypothesis is that the proportion of athletes who graduate from the university is 0.68, the all-university proportion, versus the two-sided alternative. Thus we test the hypotheses

$$H_0: p = 0.68$$
$$H_a: p \neq 0.68$$

Step 2. Test statistic. We notice that $\hat{p} = 45/74 = 0.608$. The z test statistic is therefore

$$z = \frac{\hat{p} - p_0}{\sqrt{\dfrac{p_0(1 - p_0)}{n}}} = \frac{0.608 - 0.68}{\sqrt{\dfrac{0.68(1 - 0.68)}{74}}} = \frac{-0.072}{0.054} = 1.33$$

Step 3. P-value. Because the test is two-sided, the *P*-value is the area under a standard normal curve more than 1.33 away from 0 in either direction. The figure below shows this area.

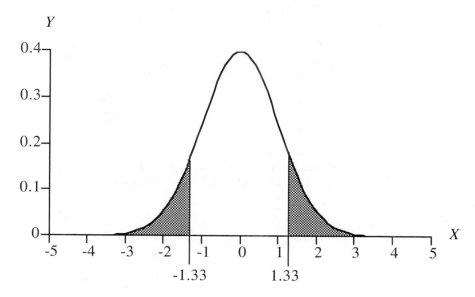

From Table A we find that the area below −1.33 is 0.0918. The *P*-value is twice this area.

$$P = 2(0.0918) = 0.1836$$

Conclusion. The data here do not provide strong evidence that the proportion of athletes that graduate within six years from this university differs from the all-university proportion of 0.68.

(b) *Step 1: Hypotheses.* Let female athletes be population 1 and male athletes population 2. Let p_1 and p_2 be, respectively, the proportion of all female and male athletes the university will admit under its present standards who will graduate within six years. We wish to see if a smaller proportion of male athletes than female athletes graduate within six years. The hypotheses we wish to test are therefore

$$H_0: p_1 = p_2$$
$$H_a: p_1 > p_2$$

Step 2. Test statistic. The pooled proportion of female and male graduates who graduated within six years iss

$$\hat{p} = \frac{\text{count of all female and male athletes who graduate within six years}}{\text{count of observations in both samples combined}}$$

$$= \frac{45}{74} = 0.608$$

The z test statistic is

$$z = \frac{\hat{p}_1 - \hat{p}_2}{\sqrt{\hat{p}(1-\hat{p})\left(\dfrac{1}{n_1} + \dfrac{1}{n_2}\right)}}$$

$$= \frac{21/28 - 24/46}{\sqrt{0.608(1-0.608)\left(\dfrac{1}{28} + \dfrac{1}{46}\right)}}$$

$$= \frac{0.228}{0.117}$$

$$Z = 1.95$$

Because the test is one-sided, the P-value is the area under a standard normal curve above 1.95. The figure below shows this area.

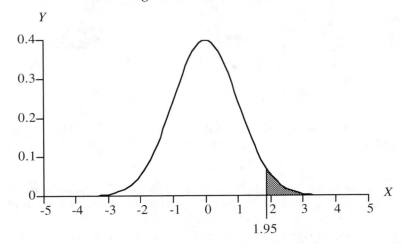

From Table A we find that the area above 1.95 is $1 - 0.9744 = 0.0256$. The P-value is thus

$$P = 0.0256$$

Conclusion. We conclude that the data provides fairly strong evidence that a smaller proportion of male athletes than female athletes graduate within 6 years.

(c) In part (a), the null hypothesis gives the value of $p_0 = 0.68$. If we regard the athletes admitted in a specific year as an SRS from the large population of athletes the university will admit under its present standards, this large population of athletes the university will admit under its present standards population size is certainly at least ten times the sample size. We also note that $np_0 = 74(0.68) = 50.32$ and $n(1 - p_0) = 74(1 - 0.68) = 23.68$, both of which are larger than 10, so it is safe to use the procedures of this chapter in part (a).

The sample sizes are, for female athletes $n_1 = 28$ and for male athletes $n_2 = 46$. The sample proportions are $\hat{p}_1 = 21/28 = 0.75$ and $\hat{p}_2 = 24/46 = 0.522$. If we regard the athletes admitted in a specific year as an SRS from the large population of athletes the university will admit under its present standards, this large population of athletes the university will admit under its present standards population size is certainly at least ten.

We note that $n_1 \hat{p}_1 = 28(0.75) = 21$, $n_1(1 - \hat{p}_1) = 28(1 - 0.75) = 7$, $n_2 \hat{p}_2 = 46(0.522) = 24$, and $n_2(1 - \hat{p}_2) = 46(1 - 0.522) = 22$. All are at least 5, so it would appear safe to use the z test.

For the baseball players, $n = 5$ and $\hat{p} = 0.6$. Both $n\hat{p} = 3$ nor $n(1 - \hat{p}) = 2$ are smaller than 5. It is not safe to use the z test.

CASE STUDY

Many people are fascinated by the paranormal. TV shows such as *Unsolved Mysteries*, *Sightings*, and the *X-Files* attest to this fact. There are even research journals, the *Journal of Parapsychology* and the *Journal of the American Society for Psychical Research* are devoted to the topic. You can probably find these in your college library. While many people are convinced of the existence of the paranormal, skeptics point to the lack of convincing, repeatable experimental evidence. In an attempt to provide such evidence, three researchers, N. S. Don, B. E. McDonough, and C. A. Warren, conducted an experiment with a "controversial" psychic, Olof Jonsson. They claim to have eliminated the possibility of trickery by Jonsson and report his ability to have exceeded that dictated by chance.

Did this experiment really provide good evidence of psychic powers? The best way to assess the quality of the evidence is to carefully examine the data ourselves. This includes understanding how the data were collected. Following is a description of how the experiment was run, followed by the overall results of the experiment. All this information comes from an article by the researchers. The reference is given at the end of this case study. We encourage you to track the article down in your college library as it contains more information than we can provide here.

The experiment was conducted using an Apple IIGS® computer, a video-game paddle, an internal random number generator, and a modified version of a computer program known as the *ESPerciser* ™. For each trial, the *ESPerciser* ™ program began with a screen with four blank rectangles and the words

"IMPRESSION PERIOD - PRESS BUTTON WHEN READY."

After clicking the paddle button, the program would display four black and white symbols (star, wave, cross, and circle) in a random order on the center of the screen. The symbols used are four of the five traditional ESP symbols. This screen was followed by another screen containing all four symbols and the words

"USE PADDLE TO POINT, PRESS BUTTON TO SELECT."

After Jonsson made his selection, the program asked him to indicate the level of his confidence in his selection as low, medium, or high. If Jonsson's guess matched the symbol selected by the computer, the word

"HIT"

was repeatedly flashed in large multicolored letters, along with "rewarding auditory sound effects."

The experiment was conducted in three sessions on three consecutive days. Each session consisted of four runs of 24 trials for a total of 96 trials each day and 288 trials in all? For each run (an entire set of 24 trials) the computer was in either a precognition mode or a clairvoyance mode. The mode was determined at random by the computer, with both modes having an equal chance of being selected. In the precognition mode, the computer selected the "correct" symbol at random *after* Jonsson made his guess. This mode presumably tests one's ability to predict the future. In the clairvoyance mode, the computer randomly selected the "correct" symbol immediately prior to the message "IMPRESSION PERIOD - PRESS BUTTON WHEN READY. This mode presumably tests traditional ESP. Jonsson was not informed of which mode the computer was in until after the entire run was completed. The researchers pointed out that, in fact, the computer selected the clairvoyance mode in 9 of the 12 runs. Was this unusual? As we have seen in previous chapters of the text, randomness does not imply perfectly uniform results. Nine out of 12 is not an extremely rare outcome. To see this, try flipping a coin 12 times and keeping track of the number of heads and tails. Do this many times. How many times do you get as many as 9 heads or 9 tails?

The researchers took elaborate precautions to prevent Jonsson from cheating. He was never left alone with the computer or software so as to prevent tampering. The random number generator used by the software was carefully checked to verify that it was really random. Randomness was used at several stages of the experiment (in the selection of the mode, in the presentation of the order of the symbols each time to Jonsson, and in the computer's selection of the "correct" symbol). The use of the *ESPerciser* ™ eliminated the possibility of a human accomplice. Several other precautions were also taken and the researchers make a convincing argument that the results of the experiment were not due to cheating.

What do the data show? The data provided by the researchers are listed below.

	Trials	Hits	Proportion correct
All Trials	288	88	0.3056
High confidence	165	55	0.3333
Medium confidence	48	12	0.2500
Low confidence	75	21	0.2800
Precognition	72	27	0.3750
Clairvoyance	216	61	0.2824

If Jonsson was just guessing, the researchers claim that the fraction of correct selections should be 0.25, since there were four choices at each trial and only one was correct. They claim that a larger proportion of correct selections would be evidence of psychic ability. If we let p represent the probability of a correct selection (or the proportion of correct selections by Jonsson in a very long sequence of trials), we might test the researchers' hypotheses by testing

$$H_0: p = 0.25$$
$$H_a: p > 0.25$$

This can be done using the methods of this chapter, after verifying that the methods are appropriate in this case. We leave this as an exercise for you, the reader, to try. Another approach might be to construct confidence intervals for the value of p. We adopt the latter approach and calculate both 95% and 99% confidence intervals for p for the overall proportion of correct guesses, the proportion when Jonsson had a high level of confidence in his selection, and in the precognition mode (the mode in which Jonsson seemed to have performed best). The results, using the methods of this chapter, are as follows. (You might try and see if you can duplicate these results.)

	95%	99%
All trials	.3056 ± .0532 = (.2524, .3588)	.3056 ± .0700 = (.2356, .3756)
High confidence	.3333 ± .0946 = (.2387, .4279)	.3333 ± .0719 = (.2614, .4052)
Precognition	.3750 ± .1118 = (.2632, .3868)	.3750 ± .1472 = (.2278, .4222)

The one-sided hypothesis tests for these 3 cases all have P-values between .01 and .02.

These results provide moderately strong evidence that Jonsson's probability of a correct selection was larger than 0.25, particularly when he was confident of his response. In addition, the data seem to suggest Jonsson has precognition (ability to predict the future) rather than "traditional" ESP. Is this finally the solid experimental evidence that researchers have been looking for? Will this evidence convince the skeptic?

On the surface, the evidence is intriguing. Of course, stronger evidence, say P-values below .01, would have been more convincing. More trials (larger sample sizes) might have accomplished this. We would also like to have the complete data rather than the summary table given. The complete data would indicate what the selection was on each trial and what the correct answer was for each trial. The complete data would help us more carefully assess the results. In addition, we would like to actually examine the *ESPerciser*™ so we can better assess this aspect of the experiment.

Are there any objective reasons to question whether the experiment provides evidence of psychic ability? Nothing in the description of the experiment or the data is blatantly suspicious. However there is one odd feature of the experiment that raises an objection that is not clearly addressed by the researchers. We find it odd that runs are always sets of 24 trials. Why 24 rather than a round number like 10, 20, or 25? If the *ESPerciser* ™ simply chooses one of the four symbols at random on each trial, there is no clear reason why runs should consist of 24 trials. However, there is another possibility. Standard ESP decks with four symbols often consist of 24 cards, six cards containing each symbol. The deck is shuffled and then cards are selected one at a time, and the subject is asked to guess the symbol on the card. Selected cards are *not* returned to the deck. Is this how the *ESPerciser* ™ works and is this why runs consist of 24 trials? It is possible (and not clear from the description of the *ESPerciser* ™) that the randomness in the software acts like the shuffling of a deck followed by a sequence of 24 draws of cards without replacement. If this is the case, then it is possible for someone with no psychic ability but with a good memory to give more than 25% correct guesses. If you can keep track of which symbol has appeared least often so far, simply guess that symbol. To illustrate how this works, consider a deck with two cards, one black and one red. On first glance you might expect that you should be correct about half the time. Actually, you can do better. On the first draw you have a 50% chance of guessing the color correctly. Once you find out the color of the first card, though, (because you guessed correctly and the *ESPerciser* ™ flashed "HIT", or because you guessed incorrectly and nothing happened) you *know* the color of the remaining card and will be able to guess it correctly. Thus you expect to have two correct guesses half the time (the half of the times you get the first card correct) and one correct guess the other half of the time (the half of the times you guess incorrectly on the first card). On average, you get 1.5 correct out of 2 and so are correct 3/4 of the time rather than half the time. And this is without any psychic ability. A person with a good memory can carry out a similar strategy with a larger deck and more symbols. If Jonsson found out the correct answer after each of his responses, it is possible to show that with a perfect memory he should be able to get at least 32.4% correct. With imperfect memory or by only finding out the correct answer when he guessed correctly (because the *ESPerciser* ™ flashed "HIT", but also note that Jonsson would know at least which symbol was not correct when he was wrong) Jonsson would generally not do as well as 32.4% correct, but would still do better than 25% correct *without having any psychic ability*. We note that Jonsson's overall results (approximately 30% correct) are consistent with this explanation.

Have we "debunked" the experiment? Not necessarily. Unfortunately, the description of the experiment does not make it clear whether the *ESPerciser* ™ randomly selected the correct response independently on each trial or whether it simulated shuffling a deck of 24 cards followed by selection without replacement. We would like to examine the version of the *ESPerciser* ™ used by the researchers in order to determine how it actually worked. In addition, if we had the complete data we could see if Jonsson was following some sort of guessing strategy, perhaps like the one we described above. Once again we encounter the same old lesson. We need *all* the data from and *all* the information about a study in order to assess the quality of the results!

Reference: Don, N. S., McDonough, B. E., and Warren, C. A. (1992). "Psi testing of a controversial psychic under controlled conditions," *Journal of Parapsychology* **56**, 87-96.

CHAPTER 8

EXAMINING RELATIONSHIPS

CHAPTER OVERVIEW

In the first chapters on inference we have looked at the ways to draw conclusions about a single population or about two populations. Now we get to the part about more than two populations or treatment groups in an experiment. Chapter 8 will show us the inference methods for more than two population proportions. When comparing more than two populations, don't just repeat the methods for comparing two populations many times. The method of this chapter is an overall test to show if there is any difference among the populations. As before, we will use subscripts to tell which statistics and parameters belong to each population.

Inference for more than two population proportions uses a **two-way table**. Two-way tables were introduced in Chapter 2, Section 2.5; the tables are a way of displaying the relationship between any two categorical variables. We will compare the **observed counts** with the counts we expect (**expected counts**) to get if the null hypothesis, $H_0: p_1 = p_2 = \ldots = p_n$, which is saying that all the population proportions are the same, is true. A more general statement of the null hypothesis is that it is saying there is no relationship between the row variable and the column variable. The new statistic is the **chi-square statistic**, and has a distribution from one of the family of chi-square distributions defined by the degrees of freedom.

KEY CONCEPTS

Making a two-way table

In this chapter the two-way tables have as many rows as there are populations or treatment groups (for full details on two-way tables see Section 2.5). The columns represent the successes and failures. Record each observation in the row which it belongs as a success or failure. The tables are also called $r \times c$ tables, where r is the number of rows and c is the number of columns.

Calculate the row columns, the column totals, and the overall table total.

The chi-square distribution

- The distribution has only positive values.
- The shape of the distribution has a long right-hand tail and is very much skewed to the right.
- The distribution is completely defined by the degrees of freedom, which are $(r-1)(c-1)$. r is the number of rows, and c is the number of columns.
- P-values represent the area to the right of the statistic under the density curve.

The chi-square test

We will use the chi-square statistic, X^2, to measure the difference between populations. The statistic is

$$X^2 = \sum \frac{(\text{observed count} - \text{expected count})^2}{\text{expected count}}$$

1. Fill in the two-way table, including the row and column totals.

2. Find the expected count for each cell in the table. The expected count is found as follows: $\dfrac{\text{row total} \times \text{column total}}{\text{table total}}$

3. Calculate the X^2 statistic for the table.

4. The test always has the hypotheses $H_0: p_1 = p_2 = \ldots = p_n$ and H_a: not all the population proportions are the same.

5. The critical values and P-values are found in Table E. Look up the values for the chi-square distribution with $(r-1)(c-1)$ degrees of freedom.

Assumptions for the chi-square test

- The data are independent SRSs from several populations and each observation is classified according to one categorical variable.
- The data are from a single SRS and each observation is classified according to two categorical variables.
- The data are an entire population classified according to two categorical variables.

- No more than 20% of the cells in the two-way table have expected counts of at least 5.

- All cells have an expected count of at least 1.

ANSWERS TO SELECTED TEXT EXERCISES

Exercise 8.1

(a) There are $r = 2$ rows ("C or better" and "D or F") and $c = 3$ columns ("< 2," "2 to 12," and "> 12").

(b) If we add the counts (students) in each of the three groups we find

	≤ 2	2 to 12	> 12
C or better	11	68	3
D or F	9	23	5
Totals	20	91	8

From the group (column) totals, we calculate the proportion of successful students in each column.

	≤ 2	2 to 12	> 12
C or better	11/20 = .550	68/91 = .747	3/8 = .375

The proportions seem to suggest that spending too much time in extracurricular activities is associated with a lack of success. Spending little time in extracurricular activities seems to be slightly associated with success. Spending a moderate amount of time appears to be the group most strongly associated with success. Perhaps the maxim "moderation in all things" applies here!

(c) Here is a bar chart of the percentages in (b).

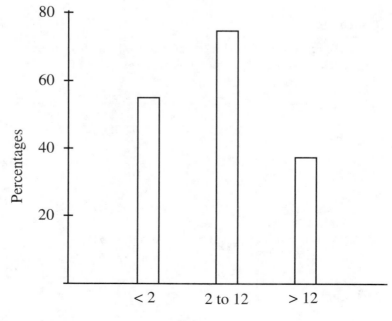

(d) If the null hypothesis is true, then the proportion of expected counts should be the same for both "C or better" and "D or F" students in all three groups. Since there are a total of 119 students, the total proportion in each group is

	≤ 2	2 to 12	≥ 12
Total proportion	20/119 = .168	91/119 = .765	8/119 = .067

These are the proportions of the 82 "C or better" students and the proportions of the "D or F" students that should be in each group if the null hypothesis is true. Converting these proportions to counts out of 82 (for the "C or better" students) and out of 37 (for the "D or F" students) respectively, yields the following table of expected counts.

	≤ 2	2 to 12	≥ 12
C or better	.168(82) = 13.78	.765(82) = 62.73	067(82) = 5.49
D or F	.168(37) = 6.22	.765(37) = 28.31	067(37) = 2.48

(e) Here is a table presenting the observed and expected counts together for successful students.

	≤ 2	2 to 12	≥ 12
C or better	Observed = 11	Observed = 68	Observed = 3
	Expected = 13.78	Expected = 62.73	Expected = 5.49
D or F	Observed = 9	Observed = 23	Observed = 5
	Expected = 6.22	Expected = 28.31	Expected = 2.48

None of the deviations are enormous, but neither is there extremely close agreement between the observed and expected values. The largest deviations between the observed and expected counts (in an absolute sense) occur for the 2 to 12 hours in the extracurricular activities group. The observed number of successful students is larger than expected and the observed number of unsuccessful students is smaller than expected. This may suggest an association between success in schoolwork and a moderate involvement in extracurricular activities.

Exercise 8.3

(a) We reproduce the Minitab output from the text below.
Expected counts are printed below observed counts

	C1	C2	C3	Total
1	11	68	3	82
	13.78	62.71	5.51	
2	9	23	5	37
	6.22	28.29	2.49	
Total	20	91	8	119

ChiSq = 0.561 + 0.447 + 1.145 + 1.244 + 0.991 + 2.538 = 6.926
df = 2
1 cells with expected counts less than 5.0
Chi-square 2.
6.9260   0.9687

The components of the chi-square statistic are calculated from the expected and observed count in each cell. They are

cell in row 1, column 1: $\dfrac{(\text{observed count} - \text{expected count})^2}{\text{expected count}} = \dfrac{(11 - 13.78)^2}{13.78} = 0.561$

cell in row 1, column 2: $\dfrac{(\text{observed count} - \text{expected count})^2}{\text{expected count}} = \dfrac{(68 - 62.71)^2}{62.71} = 0.446$

cell in row 1, column 3: $\dfrac{(\text{observed count} - \text{expected count})^2}{\text{expected count}} = \dfrac{(3 - 5.51)^2}{5.51} = 1.143$

cell in row 2, column 1: $\dfrac{(\text{observed count} - \text{expected count})^2}{\text{expected count}} = \dfrac{(9 - 6.22)^2}{6.22} = 1.243$

cell in row 2, column 2: $\dfrac{(\text{observed count} - \text{expected count})^2}{\text{expected count}} = \dfrac{(23 - 28.29)^2}{28.29} = 0.989$

cell in row 2, column 3: $\dfrac{(\text{observed count} - \text{expected count})^2}{\text{expected count}} = \dfrac{(5 - 2.49)^2}{2.49} = 2.530$

We calculate the value of the X^2 statistic by summing these 6 components. The sum is

$$X^2 = 6.912$$

The above results are quite close to, but not exactly equal to, the values on the Minitab output. This is undoubtedly because the printed output rounds off the expected counts to two decimal places, but the subsequent calculations use "unrounded" values. Since our calculations are based on the "rounded off" values, they do not exactly agree with the values on the Minitab output.

(b) The probability of obtaining a value of the X^2 statistic equal to or smaller than the observed value (6.926 on the Minitab output) is given at the bottom of the Minitab output as 0.9687. Thus the P-value (the probability of obtaining a value of the X^2 statistic as large or larger than observed) is

$$P\text{-value} = 1 - 0.9687 = 0.0313$$

We would reject H_0 at level 0.05 but not at level 0.01. Rejecting H_0 means that the observed differences in the proportions in each group for the successful ("C or better") and unsuccessful students ("D or F") cannot be easily attributed to chance. In other words, there is evidence that the proportions of successful students in the three groups (representing amounts of time spent on extracurricular activities) are different from the proportions of unsuccessful students in these groups.

(c) The term contributing most to X^2 is the term for the number of unsuccessful students spending more than 12 hours per week in extracurricular activities. This points to the fact that if one spends too much time in extracurricular activities, one has little time for schoolwork and is thus likely to be unsuccessful.

(d) The study was not a designed experiment and hence does not prove that spending more or less time on extracurricular activities *causes* changes in academic success. While at first glance it seems plausible that changing the amount of time spent on extracurricular activities ought to cause changes in academic success, this assumes (among other things) that students will trade time spent on schoolwork with time spent in extracurricular activities. However, some students may seek out involvement in extracurricular activities as an escape from schoolwork. If they aren't involved in extracurricular activities, they will simply "goof off." Changes in extracurricular activity thus will not necessarily produce changes in academic performance.

Exercise 8.5

(a) As we saw in Exercise 8.1(a), there are $r = 2$ rows ("C or better" and "D or F") and $c = 3$ columns ("< 2," "2 to 12," and "> 12"). Thus the number of degrees of freedom is

$$(r - 1)(c - 1) = (2 - 1)(3 - 1) = (1)(2) = 2$$

which agrees with the value in the Minitab output (see Exercise 8.3 above).

(b) If we look in the $df = 2$ row of Table E, we find the following information.

p	.05	.025
x^*	5.99	7.38

$X^2 = 6.926$ lies between the entries for $p = .05$ and $p = .025$. This tells us that the P-value is between .05 and .025.

Exercise 8.7

(a) We convert the percentages given to counts, rounding off to the nearest integer. Here is the resulting 4×2 table. Notice that the number of failures is obtained by subtracting the number of successes from the total number of students in the row category.

	Success	Failure
C or better	$(.36)(14) = 5$	$14 - 5 = 9$
B or better	$(.64)(64) = 41$	$64 - 41 = 23$
A	$(.90)(30) = 27$	$30 - 27 = 3$
Creative work beyond an A	$(.82)(11) = 9$	$11 - 9 = 2$

(b) The null hypothesis of the study would appear to be that the probability of success is the same regardless of a student's goals. Thus to calculate the expected counts, we obtain the column totals, which are

$$\text{total successes} = 5 + 41 + 27 + 9 = 82$$
$$\text{total failures} = 9 + 23 + 3 + 2 = 37$$

The proportions of successes and failures overall are then

$$\text{overall proportion of successes} = 82/119 = 0.689$$
$$\text{overall proportion of failures} = 37/119 = 0.311$$

We obtain the expected count in any "success" cell by multiplying the total number of students in the row by the overall proportion of successes. Likewise we obtain the expected count in any "failure" cell by multiplying the total number of students in the row by the overall proportion of failures. The results are

	Success	Failure
C or better	$(.689)(14) = 9.646$	$(.311)(14) = 4.354$
B or better	$(.689)(64) = 44.096$	$(.311)(64) = 19.904$
A	$(.689)(30) = 20.670$	$(.311)(30) = 9.330$
Creative work beyond an A	$(.689)(11) = 7.579$	$(.311)(11) = 3.421$

We notice that 2 out of 8, or 25%, of the cells have expected counts below 5. This is larger than the "safe" percentage of 20%.

(c) Since there are $r = 4$ rows and $c = 2$ columns in our table, the degrees of freedom for the chi-square statistic are

$$(r - 1)(c - 1) = (4 - 1)(2 - 1) = (3)(1) = 3$$

To get an approximate P-value we need to locate about where in row 3 of Table E a value of $X^2 = 14.986$ falls. Here is the relevant portion of Table E.

p	.0025	.001
x^*	14.32	16.27

The P-value is thus between .0025 and .001.

(d) We reproduce our 4×2 table, with both the observed and expected counts.

	Success	Failure
C or better	observed = 5	observed = 9
	expected = 9.646	expected = 4.354
B or better	observed = 41	observed = 23
	expected = 44.096	expected = 19.904
A	observed = 27	observed = 3
	expected = 20.670	expected = 9.330
creative work beyond an A	observed = 9	observed = 2
	expected = 7.579	expected = 3.421

We notice that the deviations from the expected number of successes are negative for the "C or better" and "B or better" rows, while the deviations are positive for the "A" and "creative work beyond an A" rows. This indicates there is an association between a student's expectations and her/his probability of success. Higher expectations are associated with a greater chance of success.

Exercise 8.17

We wish to see if the proportion responding "Yes" in each of the three groups is the same. Formally we are testing

> H_0: there is no relation between the response of an individual and the method by which they were asked the question.

The data given in the problem can be put into the following 3×2 table. The percentages responding "Yes" are converted to counts by multiplying them by the number of people (800 in each case) contacted by the given method. The number of "No" responses is determined by subtracting the number of "Yes" responses from the total number of people (800) contacted by the given method. The results are

	Yes	No
Phone	(.21)(800) = 168	800 − 168 = 632
One-on-one	(.25)(800) = 200	800 − 200 = 600
Anonymous written	(.28)(800) = 224	800 − 224 = 576

To determine the test statistic, we also need to compute the expected number of counts in each cell, assuming H_0 is true. These values are determined by first calculating the overall proportions of "Yes" and "No" responses out of all 2400 participants. These values are

overall proportion of "Yes"

$$= \text{total number of "Yes" responses}/2400$$
$$= (168 + 200 + 224)/2400$$
$$= 592/2400$$
$$= 0.247$$

overall proportion of "No"

$$= \text{total number of "No" responses}/2400$$
$$= (632 + 600 + 576)/2400$$
$$= 1808/2400$$
$$= 0.753$$

The table of expected counts is then

	Yes	No
Phone	(.247)(800) = 197.6	(.753)(800) = 602.4
One-on-one	(.247)(800) = 197.6	(.753)(800) = 602.4
Anonymous written	(.247)(800) = 197.6	(.753)(800) = 602.4

We combine our tables of observed and expected counts into a single table for ease of reference.

	Yes	No
Phone	observed = 168	observed = 632
	expected = 197.6	expected = 602.4
One-on-one	observed = 200	observed = 600
	expected = 197.6	expected = 602.4
Anonymous written	observed = 224	observed = 576
	expected = 197.6	expected = 602.4

For each cell we calculate the component it contributes to X^2.

cell in row 1, column 1: $\dfrac{(\text{observed count} - \text{expected count})^2}{\text{expected count}} = \dfrac{(168 - 197.6)^2}{197.6} = 4.434$

cell in row 1, column 2: $\dfrac{(\text{observed count} - \text{expected count})^2}{\text{expected count}} = \dfrac{(632 - 602.4)^2}{602.4} = 1.454$

cell in row 2, column 1: $\dfrac{(\text{observed count} - \text{expected count})^2}{\text{expected count}} = \dfrac{(200 - 197.6)^2}{197.6} = 0.029$

cell in row 2, column 2: $\dfrac{(\text{observed count} - \text{expected count})^2}{\text{expected count}} = \dfrac{(600 - 602.4)^2}{602.4} = 0.010$

cell in row 3, column 1: $\dfrac{(\text{observed count} - \text{expected count})^2}{\text{expected count}} = \dfrac{(224 - 197.6)^2}{197.6} = 3.527$

cell in row 3, column 2: $\dfrac{(\text{observed count} - \text{expected count})^2}{\text{expected count}} = \dfrac{(576 - 602.4)^2}{602.4} = 1.157$

The value of the X^2 statistic is the sum of these six components. Adding yields

$$X^2 = 4.434 + 1.454 + 0.029 + 0.010 + 3.527 + 1.157 = 10.611$$

Since our table has $r = 3$ rows and $c = 2$ columns, the number of degrees of freedom for the chi-square statistic is

$$(r - 1)(c - 1) = (3 - 1)(2 - 1) = (2)(1) = 2$$

To determine the P-value we look in row 2 of Table E. The relevant portion of the table is

p	.005	.0025
x^*	10.60	11.98

The P-value is therefore between .005 and .0025.

We conclude that there is strong evidence that there is a relation between the response of an individual and the method by which they were asked the question. Assuming that more "Yes" responses indicate greater honesty, we note that the anonymous method produced a higher than expected number of honest ("Yes") responses. Use of the phone (which is not very private since the respondent doesn't know if others are listening or if their response is being recorded) produced a lower than expected number of honest ("Yes") responses. One-on-one interviews (which may be a bit more private than the phone, since the respondent can see who is interviewing them) produced about the expected number of honest ("Yes") responses. Thus it appears that the more "private" the methods the more likely it is that one will obtain an "honest" response to sensitive questions.

Exercise 8.23

(a) A 90% confidence interval for the proportion p of all adults who felt the military should have more control over how news organizations reported about the war uses the standard normal critical value $z^* = 1.65$ from Table C (look in the bottom row for standard normal critical values). Since the sample proportion was $\hat{p} \dfrac{527 + 96}{924 + 174} = .57$ and the sample size was $n = 924$, the confidence interval is

$$\hat{p} \pm z^* \sqrt{\frac{\hat{p}(1 - \hat{p})}{n}} = 0.570 \pm 1.65 \sqrt{\frac{.57(1 - .57)}{924}} = 0.570 \pm 1.65(.0163)$$

$$= 0.570 \pm 0.027$$

$$= (0.543, 0.597)$$

(b) We wish to see if the proportion who felt the military should have more control over how news organizations reported about the war was the same in the two populations surveyed. This could be determined by conducting a two-sample test for proportions using the methods of Chapter 7, or by a chi-square test using the methods of this chapter. We adopt the latter approach. Formally we are testing

H_0: there is no relation between the population surveyed (national vs. students) and response to the question of whether the military should have more control over how news organizations reported about the war.

The data given in the problem can be put into the following 2×2 table. The percentages responding "more control" are converted to counts by multiplying them by the number of people contacted by the given survey. The number of "no control" responses is determined by subtracting the number of "more control" responses from the total number of people contacted by the given survey. The results are

	More control	No control
National survey	$(.57)(924) = 527$	$924 - 527 = 397$
Student survey	$(.55)(174) = 96$	$174 - 96 = 78$

To determine the test statistic, we also need to compute the expected number of counts in each cell, assuming H_0 is true. These values are determined by first calculating the overall proportions favoring more control and no control out of the total number of $924 + 174 = 1098$ people in both surveys. These values are

overall proportion favoring more control
$$= \text{total number favoring more control}/1098$$
$$= (527 + 96)/1098$$
$$= 623/1098$$
$$= 0.567$$

overall proportion against more control
$$= \text{total number against more control}/1098$$
$$= (397 + 78)/1098$$
$$= 475/1098$$
$$= 0.433$$

The table of expected counts is then

	More control	No control
National survey	$(.567)(924) = 523.9$	$(.433)(924) = 400.1$
Student survey	$(.567)(174) = 98.7$	$(.433)(174) = 75.3$

We combine our tables of observed and expected counts into a single table for ease of reference.

	More control	No control
National survey	observed = 527	observed = 397
	expected = 523.9	expected = 400.1
Student survey	observed = 96	observed = 78
	expected = 98.7	expected = 75.3

For each cell we calculate the component it contributes to X^2.

cell in row 1, column 1: $\dfrac{(\text{observed count} - \text{expected count})^2}{\text{expected count}} = \dfrac{(527 - 523.9)^2}{523.9} = 0.018$

cell in row 1, column 2: $\dfrac{(\text{observed count} - \text{expected count})^2}{\text{expected count}} = \dfrac{(397 - 400.1)^2}{400.1} = 0.024$

cell in row 2, column 1: $\dfrac{(\text{observed count} - \text{expected count})^2}{\text{expected count}} = \dfrac{(96 - 98.7)^2}{98.7} = 0.074$

cell in row 2, column 2: $\dfrac{(\text{observed count} - \text{expected count})^2}{\text{expected count}} = \dfrac{(78 - 75.3)^2}{75.3} = 0.097$

The value of the X^2 statistic is the sum of these four components. Adding yields

$$X^2 = 0.018 + 0.024 + 0.074 + 0.097 = 0.213$$

Since our table has $r = 2$ rows and $c = 2$ columns, the number of degrees of freedom for the chi-square statistic is

$$(r - 1)(c - 1) = (2 - 1)(2 - 1) = (1)(1) = 1$$

To determine the P-value we look in row 1 of Table E. The relevant portion of the table is the leftmost portion of the table, namely

$$p \quad .25$$
$$x^* \quad 1.32$$

The P-value is therefore larger than 0.25. We conclude that there is no compelling evidence that there is a difference between the student response and the national response.

Note: Students with calculators that calculate X^2 directly or students using statistical software such as Minitab should compare the results they get with those obtained above. You may find slight differences due to round-off in the above calculations. We encourage you to use your calculator or software rather than do calculations by hand whenever possible.

(c) We wish to see if the proportion who felt the military should exert more control is the same for the two methods of wording the question. This could be determined by conducting a two sample test for proportions using the methods of Chapter 7, or by a chi-square test using the methods of this chapter. We adopt the latter approach. Formally we are testing

H_0: there is no relation between the way the question was worded and the proportion who agree that the military should have more control over how news organizations reported about the war.

The data given in the problem can be put into the following 2×2 table. The percentages responding "more control" are converted to counts by multiplying them by the number of people contacted by the given survey. The number of "no control" responses is determined by subtracting the number of "more control" responses from the total number of people contacted by the given survey. The results are

	More control	No control
Part (a) wording	(.55)(174) = 96	174 – 96 = 78
Part (c) wording	(.16)(199) = 32	199 – 32 = 167

To determine the test statistic, we also need to compute the expected number of counts in each cell, assuming H_0 is true. These values are determined by first calculating the overall proportions favoring more control and no control out of the total number of $174 + 199 = 373$ people in both surveys. These values are

overall proportion favoring more control
= total number favoring more control/373
= (96 + 32)/373
= 128/373
= 0.343

overall proportion against more control

$$= \text{total number against more control}/373$$
$$= (78 + 167)/373$$
$$= 245/373$$
$$= 0.657$$

The table of expected counts is then

	More control	No control
National survey	(.343)(174) = 59.7	(.657)(174) = 114.3
Student survey	(.343)(199) = 68.3	(.657)(199) = 130.7

We combine our tables of observed and expected counts into a single table for ease of reference.

	More control	No control
National survey	observed = 96	observed = 78
	expected = 59.7	expected = 114.3
Student survey	observed = 32	observed = 167
	expected = 68.3	expected = 130.7

For each cell we calculate the component it contributes to X^2.

cell in row 1, column 1: $\dfrac{(\text{observed count} - \text{expected count})^2}{\text{expected count}} = \dfrac{(96 - 59.7)^2}{59.7} = 22.072$

cell in row 1, column 2: $\dfrac{(\text{observed count} - \text{expected count})^2}{\text{expected count}} = \dfrac{(78 - 114.3)^2}{114.3} = 11.528$

cell in row 2, column 1: $\dfrac{(\text{observed count} - \text{expected count})^2}{\text{expected count}} = \dfrac{(32 - 68.3)^2}{68.3} = 19.293$

cell in row 2, column 2: $\dfrac{(\text{observed count} - \text{expected count})^2}{\text{expected count}} = \dfrac{(167 - 130.7)^2}{130.7} = 10.082$

The value of the X^2 statistic is the sum of these four components. Adding yields

$$X^2 = 22.072 + 11.528 + 19.293 + 10.082 = 62.975$$

Since our table has $r = 2$ rows and $c = 2$ columns, the number of degrees of freedom for the chi-square statistic is

$$(r - 1)(c - 1) = (2 - 1)(2 - 1) = (1)(1) = 1$$

To determine the P-value we look in row 1 of Table E. The relevant portion of the table is the rightmost portion of the table, namely

p	.0005
x^*	12.12

The P-value is therefore smaller than 0.0005. We conclude that there is strong evidence that there is a difference between the student responses to the two different wordings.

Note: Again, this probably is best done using software or a calculator that automatically does the chi-square calculations. Use these tools if you have them and compare your answers with the above direct calculations. You may find slight differences due to round-off in the above calculations.

(d) We now wish to see if the proportion responding "Yes" in each of the three groups is the same. Formally we are testing

H_0: there is no relation between the response of an individual and the survey which they participated in.

The data given in the problem can be put into the following 3×2 table. The percentages responding "Yes" are converted to counts by multiplying them by the number of people in each survey. The number of "No" responses is determined by subtracting the number of "Yes" responses from the total number of people in each survey. The results are

	Yes	No
National survey	(.76)(924) = 702	924 - 702 = 222
Student survey 1	(.67)(174) = 117	174 - 117 = 57
Student survey 2	(.65)(199) = 129	199 - 129 = 70

To determine the test statistic, we also need to compute the expected number of counts in each cell, assuming H_0 is true. These values are determined by first calculating the overall proportions of "Yes" and "No" responses out of all $924 + 174 + 199 = 1297$ people in the three surveys. These values are

overall proportion of "Yes"
$$= \text{total number of "Yes" responses}/2400$$
$$= (702 + 117 + 129)/1297$$
$$= 948/1297$$
$$= 0.731$$

overall proportion of "No"
$$= \text{total number of "No" responses}/1297$$
$$= (222 + 57 + 70)/1297$$
$$= 349/1297$$
$$= 0.269.$$

The table of expected counts is then

	Yes	No
National survey	(.731)(924) = 675.4	(.269)(924) = 248.6
Student survey 1	(.731)(174) = 127.2	(.269)(174) = 46.8
Student survey 2	(.731)(199) = 145.5	(.269)(199) = 53.5

We combine our tables of observed and expected counts into a single table for ease of reference.

	Yes	No
National survey	observed = 702 expected = 675.4	observed = 222 expected = 248.6
Student survey 1	observed = 117 expected = 127.2	observed = 57 expected = 46.8
Student survey 2	observed = 129 expected = 145.5	observed = 70 expected = 53.5

For each cell we calculate the component it contributes to X^2.

cell in row 1, column 1: $\dfrac{(\text{observed count} - \text{expected count})^2}{\text{expected count}} = \dfrac{(702 - 675.4)^2}{675.4} = 1.048$

cell in row 1, column 2: $\dfrac{(\text{observed count} - \text{expected count})^2}{\text{expected count}} = \dfrac{(222 - 248.6)^2}{248.6} = 2.846$

cell in row 2, column 1: $\dfrac{(\text{observed count} - \text{expected count})^2}{\text{expected count}} = \dfrac{(117 - 127.2)^2}{127.2} = 0.818$

cell in row 2, column 2: $\dfrac{(\text{observed count} - \text{expected count})^2}{\text{expected count}} = \dfrac{(57 - 46.8)^2}{46.8} = 2.223$

cell in row 3, column 1: $\dfrac{(\text{observed count} - \text{expected count})^2}{\text{expected count}} = \dfrac{(129 - 145.5)^2}{145.5} = 1.871$

cell in row 3, column 2: $\dfrac{(\text{observed count} - \text{expected count})^2}{\text{expected count}} = \dfrac{(70 - 53.5)^2}{53.5} = 5.089$

The value of the X^2 statistic is the sum of these 6 components. Adding yields

$$X^2 = 1.048 + 2.846 + 0.818 + 2.223 + 1.871 + 5.089 = 13.895$$

Since our table has $r = 3$ rows and $c = 2$ columns, the number of degrees of freedom for the chi-square statistic is

$$(r - 1)(c - 1) = (3 - 1)(2 - 1) = (2)(1) = 2$$

To determine the P-value we look in row 2 of Table E. The relevant portion of the table is

p	.001	.0005
x^*	13.82	15.20

The P-value is therefore between .001 and .0005.

We conclude that there is strong evidence that there is a relation between whether a respondent thought the news was censored and the population from which the respondent came.

Note: Once again, this probably is best done using software or a calculator which automatically does the chi-square calculations. Use these tools if you have them and compare your answers with the above direct calculations. You may find slight differences due to round-off in the above calculations.

CASE STUDY

Are you left-handed or right-handed? What about your pet? Is your dog a "lefty"? Most humans prefer using one hand over the other for most tasks. For a long time it was thought that this handedness, also called behavioral lateralization, was unique to humans. Modern research is showing us that humans may not be so unique after all; ancient fossils now suggest that behavioral lateralization in the animal kingdom may have existed for at least 500 million years.

Loren Babcock, a researcher at Ohio State University, looked at the fossilized remains of a group of extinct marine animals called trilobites, which are distantly related to the living horseshoe crab and other crustaceans. Trilobites lived on the sea floor in many areas and are believed to have moved about like modern crabs and lobsters. Babcock studied those trilobite fossils which showed healed injuries as indicated by "obvious physical breaks in the trilobites' crusty exoskeletons that have subsequently healed, usually by 'callusing' over." Fatal injuries could not be studied because the skeletons, if left at all, would show no signs of healing and those fatal breaks cannot be distinguished from breaks made after the trilobite died.

The healed injuries were divided into two categories: those of uncertain origin (possibly accidents during molting, copulation, or combat) and those caused by predators. The study was particularly interested in those breaks most likely caused by predators (if the injury was located on a section of the body that was unlikely to be accidentally injured, if the break covered a large area, and if they were not simple breaks that could have been accidental). Breaks obviously caused by predators' teeth and/or claws were most common on the edges, the rear of the body, and the right side of the trilobite. A preference for one side over the other, as compared to other types of non-lethal injuries might be taken as evidence of behavioral lateralization or "handedness," since these might indicate a preference on the part of predators to attack from one side rather than the other.

The table below contains the counts of the incidence of healed injuries according to origin of the injury and the location of the injury.

	Right side only	Left side only
Non-lethal injuries of uncertain origin	42	34
Non-lethal predation scars	60	23

Note: Specimens with multiple injuries on one side were only counted once.

We can use the methods of this chapter to see if the proportion of right side only injuries differ for predation scars as compared to injuries of other origins. A difference might be an indication of behavioral lateralization. We test the hypothesis

H_0: there is no difference in the proportion of right side only injuries due to predation as compared with injuries of other origins.

To test the hypothesis, we produce a revised table containing observed and expected counts.

	Right side only	Left side only
Non-lethal injuries of uncertain origin:	observed = 42 expected = 48.755	observed = 34 expected = 27.245
Non-lethal predation scars:	observed = 60 expected = 53.245	observed = 23 expected = 29.745

This leads to a chi-square statistic of $X^2 = 5.001$ having 1 degree of freedom. This X^2 statistic has a P-value of 0.0253. We invite you to verify these values. The results provide strong but not overwhelming evidence that there is a difference in the proportion of right side only injuries due to predation as compared with injuries of other origins. This is consistent with the assumption of behavioral lateralization. Does it prove that behavioral lateralization, at least in some animals, exists?

Notice that this is an observational study rather than a designed experiment. It is also that the "sample" is not really a random sample. While the results are consistent with behavioral lateralization, they do not "prove" it exists in animals. In observational studies we must always be on the alert for confounding factors. These might include the following:

• It is not clear from the protocol whether the division of trilobite fossils was done blindly by someone unaware the question being addressed. If the division was not done blindly, then the results could be biased because the researchers may have unintentionally classified the fossils one way or another in support of their theory.

• Another possibility that one must rule out is an argument like "The only trilobites making it into the study are those that became fossils. Perhaps trilobites injured on the right side were less able to move against the current and avoid getting stuck in the mud, thus they became fossils."

• The question is whether an injury on the right side is in any way associated with the chance of being in the study (i.e., becoming fossilized). If so, confounding is present.

• Sherlock Holmes said that once we have eliminated all other possibilities, what remains must be the answer. However, how do we know when we have considered "all other" possibilities? Just because we can't think of alternative explanations for the results of a study does not mean they do not exist. It may be that something we have not considered explains the observed results of this study. This is something that we must always contend with in nonrandomized observational studies.

Perhaps the safest conclusion is to treat the results of this study as intriguing evidence. The data form just one piece of the case for behavioral lateralization in animals. Data from the past, such as that used in this study, will always be nonrandomized observational data. In such cases, additional evidence is needed to make a strong case. It would be unsafe to rest the case for "handedness" of animals on this study alone. The next time you play with your pet, check to see if it is a "righty" or a "lefty."

CHAPTER 9

ANALYSIS OF VARIANCE: COMPARING SEVERAL MEANS

CHAPTER OVERVIEW

In our discussion of inference methods we have reached the point of talking about more than two population means. When the mean is the best description of the center of a distribution, we may want to compare several population means or several treatment means in a designed experiment. We may want to look at the mean weight loss by dieters on three different diet programs or the mean yield of four varieties of green beans.

The method we use to compare more than two population means is the **analysis of variance (ANOVA) F test**. This test is also called the **one-way ANOVA**. This is a different test from the F test you studied in Chapter 6 that compared the standard deviations of two populations. The test looks at the mean of each sample relative to the variation in the observations. This is an overall test that looks for any difference between the means. Our follow-up to this overall test will be to use the tools of the first chapters to decide in what ways the means differ. The null hypothesis is H_0: $\mu_1 = \mu_2 = \ldots = \mu_n$, where we can tell the population means apart by the subscripts, 1 through n, for as many populations as we have. The alternative hypothesis is H_a: not all the means are equal.

The ANOVA F test compares the variation *among* the groups to the variation *within* the groups. Another way of saying this is that we look for the variation expected due to the different populations and compare it to the variation among observations we expect to be similar. The important thing to take away from this chapter is the idea of the ANOVA F test. The particulars of the calculation are not as important, since software usually calculates the numbers for us.

KEY CONCEPTS

The *F* statistic

The *F* statistic is a ratio of variations.

$$F = \frac{\text{variation among the groups}}{\text{variation within the groups}} \quad \text{or}$$

$$= \frac{\text{variation among the sample means}}{\text{variation among the individuals}}$$

The *F* statistic has the *F* distribution. The distribution is completely defined by its two degrees of freedom parameters, the numerator degrees of freedom and the denominator degrees of freedom. The numerator has $(I - 1)$ degrees of freedom, where I is the number of populations we are comparing. The denominator has $(N - I)$ degrees of freedom, where N is the total number of observations.

- The *F* distribution is usually written $F(I - 1)(N - I)$.
- The statistic takes only positive or zero values.

Assumptions about ANOVA

- There are I independent SRSs.
- Each population is normally distributed with its own mean, μ_i.
- All populations have the same standard deviation, σ.

The first assumption is the most important. The test is robust against non normality, however it is still important to check for outliers and skewness that would make the mean a poor measure of the center of the distribution. As for the assumption of equal standard deviations, make sure that the largest sample standard deviation is no more than twice the smallest standard deviation.

Calculations for the ANOVA *F* test

Each population is distributed as $N(\mu_i, \sigma)$, with sample size n_i, sample mean $\bar{x}_i$, and sample standard deviation, s_i.

The *F* statistic is $F = \dfrac{\text{MSG}}{\text{MSE}}$, where MSG is the mean square for groups,

$$\text{MSG} = \frac{n_1(\bar{x}_1 - \bar{x})^2 + n_2(\bar{x}_2 - \bar{x})^2 + \ldots + n_I(\bar{x}_I - \bar{x})^2}{I - 1} \quad \text{and MSE is the error mean square,}$$

$$\text{MSE} = \frac{s_1^2(n_1 - 1) + s_2^2(n_2 - 1) + \ldots + s_I^2(n_I - 1)}{N - I}.$$

Compare the *F* statistic to the critical values of the *F* distribution with I-1 and N-I degrees of freedom. $\bar{x}$ is the overall mean of all the observations.

We can also make a confidence interval for any of the means by using the form $\bar{x}_i \pm t^* \dfrac{s_p}{\sqrt{n_i}}$. The critical value is t^* from the t distribution with $N - I$ degrees of freedom; $s_p = \sqrt{\text{MSE}}$.

ANSWERS TO SELECTED TEXT EXERCISES

Exercise 9.3

(a) Here are stemplots for each number of plants per acre. We round off yield to the nearest bushel per acre so as to make the stemplots more informative

12,000	16,000	20,000	24,000	28,000
11 | 3 8	11 |	11 |	11 |	11 | 9
12 |	12 | 1	12 |	12 |	12 |
13 |	13 | 5	13 | 0	13 | 5 8	13 |
14 | 3	14 |	14 | 0	14 |	14 |
15 | 0	15 | 0	15 | 0	15 | 6	15 | 1
16 |	16 | 7	16 | 5	16 |	16 |

The means for the 5 planting rates are

12,000: $\bar{x} = 131.025$

16,000: $\bar{x} = 143.150$

20,000: $\bar{x} = 146.225$

24,000: $\bar{x} = 143.067$

28,000: $\bar{x} = 134.750$

The data appear to show that for very low (12,000) and very high (28,000) planting rates the mean is lower than for intermediate (16,000, 20,000, or 24,000) rates. The yield first increases as planting rate increases, reaches a maximum at a planting rate of 20,000, and then decreases as planting rate continues to increase.

(b) Let μ_1, μ_2, μ_3, μ_4, and μ_5 represent the appropriate population means for the planting rates of 12,000, 16,000, 20,000, 24,000, and 28,000, respectively. The hypotheses for the ANOVA F test in this situation are then

H_0: $\mu_1 = \mu_2 = \mu_3 = \mu_4 = \mu_5$

H_a: not all of μ_1, μ_2, μ_3, μ_4, and μ_5 are equal.

(c) The ANOVA F test gives a value for the F statistic of $F = 0.50$. Its P-value is 0.736. This tells us that there is not strong evidence that the mean yields differ for the different planting rates.

(d) While the differences in the mean yields are large, ranging from a low of 131.025 to a maximum of 146.225, there is also considerable variation in the values for each planting rate. This variation can be seen in Figure 9.4 of the text in the standard deviations, ranging from 11.44 to 22.27, reported for each planting rate. This is also evident in our stemplots for the 5 planting rates (see part (a)). The values in the 5 stemplots overlap and show considerable variation. This variation indicates that there is considerable uncertainty in how close the sample means are to the population means, and hence we are not able to assert that there is strong evidence of a difference in these population means.

Exercise 9.5

(a) I = the number of populations we wish to compare
     = the number of different planting rates used
     = 5

n_1 = the sample size from population 1 (planting rate of 12,000) = 4
n_2 = the sample size from population 2 (planting rate of 16,000) = 4
n_3 = the sample size from population 3 (planting rate of 20,000) = 4
n_4 = the sample size from population 4 (planting rate of 24,000) = 3
n_5 = the sample size from population 5 (planting rate of 28,000) = 2

N = the total sample size = the sum of the n_i = 17

(b) The degrees of freedom for the ANOVA F statistic are

$$I - 1 = 5 - 1 = 4$$

degrees of freedom in the numerator and

$$N - I = 17 - 5 = 12$$

in the denominator. These agree with the values given in Figure 9.4 of the text.
(c) Here is the relevant information from Table D.

p	Critical value
0.100	2.48
0.050	3.26
0.025	4.12
0.010	5.41
0.001	9.63

We see that the P-value for F = 0.50 is larger than 0.100.

Exercise 9.9

(a) The most obvious feature is that men who are or have been married earn more, on average, than single men. Men who are, or have been, married earn about the same amount, although divorced men appear to earn a little less, on average, then married or widowed men.
(b) The ratio of the largest to the smallest standard deviations is

$$\frac{\text{largest sample standard deviation}}{\text{smallest sample standard deviation}} = \frac{8119}{5731} = 1.42 < 2$$

Since this ratio is less than 2, the sample standard deviations allow the use of the ANOVA F test.
(c) We calculate

I = the number of populations we wish to compare
= the number of different marital status's
= 4

n_1 = the sample size from population 1 (single men) = 337
n_2 = the sample size from population 2 (married men) = 7730
n_3 = the sample size from population 3 (divorced men) = 126
n_4 = the sample size from population 4 (widowed men) = 42
N = the total sample size = the sum of the n_i = 8235

The degrees of freedom for the ANOVA F statistic are

$$I - 1 = 4 - 1 = 3$$

degrees of freedom in the numerator and

$$N - I = 8235 - 4 = 8231$$

in the denominator.

(d) The large sample sizes (particularly for the married men) indicate that the margin of errors for the sample means will be very small, much smaller than the observed differences in the means. For example, the standard error for the mean of married men is

$$\frac{\$7159}{\sqrt{7730}} = \$81.4$$

which is very small compared to the difference of, for example, over $5000 in mean salaries with single men (you can check that the standard error for single men is $312.2, also much smaller than the difference in means of over $5000).

(e) The differences in means probably do not mean that getting married raises men's mean incomes. This is an observational study, so it is not safe to conclude that the observed differences are due to cause and effect. A likely explanation of the observed differences is that the typical single man is much younger than the typical married man. As men get older they are more likely to be married. Younger men have been in the firm less than older men and so will have lower salaries. Note this explanation may explain the small differences between men who are or have been married. Married and widowed men (as compared to divorced men) are likely to include the most senior men in the firm. The most senior men are likely to have the highest salaries.

Exercise 9.13

The means and standard deviations, from Figure 9.4 in the text are

Rate	N	Mean	Std. Dev.
12,000	4	131.03	18.09
16,000	4	143.15	19.79
20,000	4	146.23	15.07
24,000	3	143.07	11.44
28,000	2	134.75	22.27

Here $I = 5$ and $N = 17$ (see the solution to Exercise 9.5). Thus,

$$
\begin{aligned}
\text{MSE} &= \frac{(n_1 - 1)s_1^2 + (n_2 - 1)s_2^2 + \text{L} + (n_I - 1)s_I^2}{N - I} \\[2mm]
&= \frac{(4 - 1)18.09^2 + (4 - 1)19.79^2 + \text{L} + (2 - 1)22.27^2}{12} \\[2mm]
&= \frac{3595.69}{12} \\[2mm]
&= 299.64
\end{aligned}
$$

$$
\begin{aligned}
\bar{x} &= \frac{1}{N}(n_1 \bar{x}_1 + n_2 \bar{x}_2 + \ldots + n_I \bar{x}_I) \\[2mm]
&= \frac{1}{17}(4(131.03) + 4(143.15) + 4(146.23) + 3(143.07) + 2(134.75)) \\[2mm]
&= \frac{2380.35}{17} \\[2mm]
&= 140.02
\end{aligned}
$$

$$\text{MSG} = \frac{n_1(\bar{x}_1 - \bar{x})^2 + n_2(\bar{x}_2 - \bar{x})^2 + \text{L} + n_I(\bar{x}_I - \bar{x})^2}{I - 1}$$

$$= \frac{4(131.03 - 140.02)^2 + 4(143.15 - 140.02)^2 + \text{L} + 2(134.75 - 140.02)^2}{5 - 1}$$

$$= \frac{600.18}{4}$$

$$= 150.05$$

The results agree quite well with the results on the output in Figure 9.4. Differences are due to round-off in the reported means and standard deviations (that we used in our calculations) in Figure 9.4.

Exercise 9.17

We begin with some descriptive statistics.

Type	N	Means	Std. Devs	Std. Error of the Mean
Compact	26	27.769231	4.0625873	0.79674
Mid-size	20	25.750000	3.6400549	0.81394
Large	13	25.076923	2.7526211	0.76344

The means are not equal but there is not a large difference between them, particularly in light of the variability in the data as indicated by the standard deviations and standard errors of the mean. It is not clear whether the differences are statistically significant.

Here are side-by-side boxplots of the highway gas mileages. These are modified boxplots. They display outliers, if any exist, and only extend the hinges to the minimum and maximum "non-outlier" values.

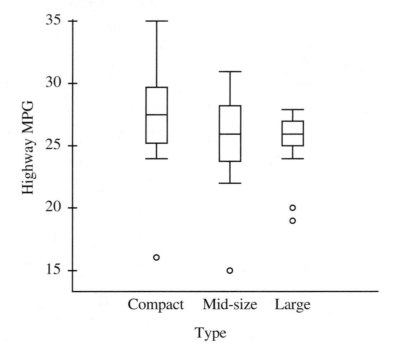

Once again, we see the medians differ, but the differences are well within the variation in the data. It is again not clear if there are significant differences in the gas mileages. We also note that there are outliers in the data for each car type. For all types of cars, the outliers are unusually low values (cars that get low gas mileages). These outliers will make the means smaller than the medians but the overall effect on each type of car should be about the same. Thus their effect on the differences in the three means is probably not large.

Before we carry out an ANOVA, we need to consider whether the assumptions for an ANOVA are plausible. First we notice that the ratio of the largest to smallest standard deviation for the three types of cars is $4.2064/2.7526 = 1.528$. This is less than 2, so there is no serious problem with the assumption of equal population standard deviations. The presence of the outliers suggests nonnormality in the data. ANOVA procedures tend to be robust to nonnormality, particular as the sample sizes get larger, and are probably safe to use here since the sample sizes are moderate. Just to check, we will run the ANOVA with the outliers present and then with the outliers removed to see if there are any marked differences. If there are, we will know that the outliers have a large effect on the results and will need to think carefully about how to interpret the data.

Analysis of Variance For Highway MPG (all data included)

Source	df	SS	MS	F	P-value
Type of car	2	79.4573	39.7287	2.9456	0.0608
Error	56	755.288	13.4873		
Total	58	834.746			

Analysis of Variance For Highway MPG (4 outliers removed)

Source	df	SS	MS	F	P-value
Type of car	2	55.2620	27.6310	3.4574	0.0389
Error	52	415.574	7.99181		
Total	54	470.836			

The P-value is smaller after removal of the outliers, but in neither case (with or without the outliers included) is there strong evidence of a difference in the mean highway gas mileages for the three types of cars.

While the sample means and medians suggest that compact cars may get a bit higher highway gas mileage than mid-sized and large cars (which get very similar mileages), there is enough variability in the data so as to make it unclear if the difference is statistically significant. An ANOVA F test produces a P-value of .0608, which is small but not unusually so. This is consistent with the story told by the descriptive statistics and graphs. We conclude that the data do not give strong evidence of a difference in the mean highway gas mileages for compact, mid-size, and large cars.

CASE STUDY

We usually think of the phrase "Let the buyer beware" when buying a used car,. However, this phrase may also apply to purchasing "diet" and "health" foods as well. Researchers have discovered that some labels on food packages understate the calorie content by more than 85%! This is not comforting news to health conscious consumers.

Four researchers[1] carried out a survey of 40 food items claiming to be "lite," "reduced-calorie," "low-calorie," "diet," "low-fat," "no-fat," or "health" foods. All items were purchased in New York City from January through August 1992. The foods were classified based on their distribution as either nationally advertised, regionally distributed, or locally prepared. "Nationally advertised" meant the food had brand names familiar to the investigators and were bought at a major super-market. Foods not made by a company the investigators recognized from national advertising were considered to be "regionally distributed." "Locally prepared" meant the food was manufactured or prepared by the vendor. The objective was to determine if the calories on the food label accurately stated the actual calories of the item and if discrepancies varied with the extent of the distribution of an item.

The actual caloric contents were measured by bomb calorimetry. This method burns the food in a controlled, small "explosion" and measures the heat generated. Four foods had labels that couldn't be removed, but in all other cases the technicians were blind to the foods' labeled caloric contents. The bomb calorimetry readings were converted into an estimate of total metabolizable energy, the type of energy the calories on a food label are supposed to reflect.

As always, in assessing the results of a study we need to look at the data. The data are given below and contain the following variables:

Food = kind of food

Difference = percentage difference between measured calories and labeled calories per item, i.e., 100% × (measured − labeled)/labeled

Ratio = labeled calories/measured calories

Classification = N if nationally advertised
R if regionally distributed
L if locally prepared

Food	Difference	Ratio	Classification
Noodles and alfredo sauce	2	0.98039216	N
Cheese curls	−28	1.3888889	N
Green beans	− 6	1.0638298	N
Mixed fruits	8	0.92592593	N
Cereal	6	0.94339623	N
Fig bars	−1	1.0101010	N
Oatmeal raisin cookie	10	0.90909091	N
Crumb cake	13	0.88495575	N
Crackers	15	0.86956522	N
Blue cheese dressing	− 4	1.0416667	N
Imperial chicken	− 4	1.0416667	N
Vegetable soup	−18	1.2195122	N
Cheese	10	0.90909091	N
Chocolate pudding	5	0.95238095	N
Sausage biscuit	3	0.97087379	N
Lasagna	−7	1.0752688	N
Spread cheese	3	0.97087379	N
Lentil soup	− 0.5	1.0050251	N
Pasta with shrimp and tomato sauce	−10	1.1111111	N
Chocolate mousse	6	0.94339623	N
Meatless sandwich	41	0.70921986	R
Oatmeal cookie	46	0.68493151	R

[1] Allison, D., Heshka, S., Sepulveda, D., and Heymsfield, S. (1993),"Counting Calories — Caveat Emptor," *JAMA* , 270, 1454–1456.

Lemon pound cake	2	0.98039216	R
Banana cake	25	0.80000000	R
Brownie	39	0.71942446	R
Butterscotch bar	16.5	0.85836910	R
Blondie	17	0.85470085	R
Oat bran snack bar	28	0.78125000	R
Granola bar	−3	1.0309278	R
Apricot bar	14	0.87719298	R
Chocolate chip cookie	34	0.74626866	R
Carrot muffin	42	0.70422535	R
Chinese chicken	15	0.86956522	L
Gyoza	60	0.62500000	L
Jelly diet candy-reds flavor	250	0.28571429	L
Jelly diet candy-fruit flavor	145	0.40816327	L
Florentine manicotti	6	0.94339623	L
Egg foo young	80	0.55555556	L
Hummus with salad	95	0.51282051	L
Baba ghanoush with salad	3	0.97087379	L

Do "health" or "diet" foods accurately describe their benefits or do they tend to overstate them? In particular, is the calorie information on their labels accurate or do manufacturers tend to understate the calorie content? Of course, manufacturers cannot make any claims they want. Labels are supposed to provide accurate information. Conscientious consumers and consumer groups are often on the alert for false information. Nationally recognized brands, by virtue of their visibility, are likely to be tested by such groups for false claims. Local or regional brands, since they are less visible, are less likely to be scrutinized by concerned consumers and hence may be more likely to "stretch" the truth. We might investigate these issues in a three step process. First, we look at the data graphically to get an overall sense of what the data are saying. Next, we might test to see if their is any difference between locally distributed, regionally distributed, and nationally distributed "health" or "diet" foods in terms of the difference between the calories in the items as listed on the labels and the calories in the items as measured in the lab. Perhaps one-way analysis of variance can be employed to test this. Our preliminary graphical exploration of the data should alert us to outliers or skewness that would make it unsafe to use one-way analysis of variance. Finally, we can estimate the mean difference between the calories on the label and the calories as measured in the lab for each population of items (locally distributed, regionally distributed, and nationally distributed items) by means of confidence intervals. These intervals will help us decide which population means differ (if the ANOVA suggests that the population means differ) and, in addition, which ones have means that indicate the label values are lower than the measured values.

One question that we must consider before proceeding is how to best measure the difference between the number of calories as indicated on an item's label and the number of calories as measured in the lab. The obvious choice is simply to use the difference in these numbers, i.e., use

measured value − label value

Notice this measure ignores the fact that a 1 calorie difference in a low calorie food item is a bigger discrepancy than a 1 calorie difference in a high calorie item. We could standardize the measure, thus fixing this problem, by looking at the percentage difference relative to the label value, i.e., we could use

100% × (measured value − label value)/(label value)

This is the measure the researchers chose and we have given the values of this variable above (see the variable difference). However when this measure is used, the data looks nonnormal and the standard deviations for the three groups differ greatly. The ratio of the largest to smallest standard deviation is much larger than 2 and so it is not safe to use ANOVA. You are invited to verify these statements by making appropriate plots of the variable difference. It turns out that for purposes of analysis it is better to use the ratio

$$(label\ value)/(measured\ value)$$

Plots (histograms, stemplots, or boxplots) of the values of this measure look reasonably normal and the ratio of the largest to smallest standard deviation is only slightly larger than 2. This measure comes closer to satisfying the conditions for safe use of one-way analysis of variance than the others. Values of this ratio that are less than 1 indicate the labeled value is less than the number of calories measured in the lab. Values greater than 1 indicate the opposite. Here are side by side boxplots of the variable ratio.

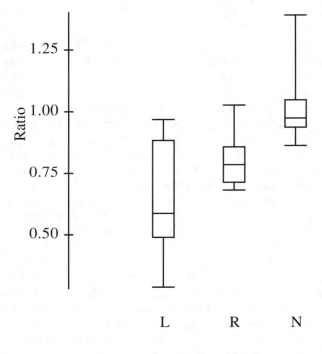

Notice that the medians for the three classifications differ. Nationally distributed items appear to have a higher value of the ratio than the other two. In addition, local and regional items appear to have values of this ratio consistently less than 1 (indicating a systematic tendency to understate the number of calories). Nationally distributed items generally appear to have ratios near or above 1. While the variability in the ratio appears different for the three classifications, the ratio of the maximum to the minimum standard deviation is only slightly larger than 2. This violates our rule of thumb only slightly and it is probably reasonably safe to proceed with a one-way ANOVA.

Let μ_1, μ_2, and μ_3 represent the mean value of this ratio for the population of locally distributed, regionally distributed, and nationally distributed "health" or "diet" items, respectively. To determine whether these means differ, we test the hypotheses

$H_0: \mu_1 = \mu_2 = \mu_3$

$H_a:$ the μ are not all the same

The new variable "ratio" will be less than 1 if the labeled understates the true calories in an item. The smaller the value of "ratio" the more seriously the labeled value understates the actual calories in an item.

The results of a one-way ANOVA are given below, followed by 95% confidence intervals for the mean for each population (locally distributed, regionally distributed, or nationally distributed). These results were obtained using statistical software.

Analysis of variance for ratio

Source	df	Sums of Squares	Mean Square	F-ratio	P-value
Classification	2	0.834030	0.417015	17.548	≤ 0.0001
Error	37	0.879283	0.023764		
Total	39	1.71331			

95% confidence intervals

mean of ratio for locally distributed items = (0.43, 0.86)
mean of ratio for regionally distributed items = (0.74, 0.88)
mean of ratio for nationally distributed items = (0.95, 1.07)

The analysis indicates a P-value of less than 0.0001 for the null hypothesis of no difference in the means. Thus there is strong evidence of a difference in the population means. The confidence intervals confirm what appears in the boxplots. Locally (relative to New York City) distributed items and regionally (relative to New York City) distributed items seem to consistently understate the calories in items (as indicated by ratios less than 1). Nationally distributed items show no such tendency (the confidence interval contains the value 1).

What should we conclude? *Assuming* the data are samples from three populations, if you live in the New York City area and are concerned about the number of calories in "health" or "diet" food items, stick to nationally distributed items. Label information appears reliable. If you do not live in or near New York City, the data do not give information about the reliability of label information concerning calories for locally or regionally distributed items in your area. Our conclusions about nationally distributed items should be valid, however. The pessimist should probably stick to national brands.

Of course, if the data were not a random sample it is at best suggestive. We cannot be sure to what extent the patterns in our boxplots extend to items other than those in the sample. It would be irresponsible to make sweeping statements about the reliability of label information of locally and regionally distributed "health" or "diet" items in general.

In closing, we would like to point out that we are now marketing locally a great diet chocolate eclair with only 5 calories! Interested?

CHAPTER 10

INFERENCE FOR REGRESSION

CHAPTER OVERVIEW

A long time ago, back in Chapter 2, we learned about regression and the idea of a **least-squares line**. When there seems to be a linear relationship between two quantitative variables, we can use **regression** and the line to both describe the connection between the variables and to make predictions about the response variable, y. In this final chapter we will see how to conduct **significance tests** about the parameters in the least-squares regression line and make **confidence intervals** for our predictions.

Before doing any inference we need to check a few things:

1. Check the scatterplot of the data. Make sure the relationship is indeed linear. (Chapter 2 has the details on how to create a scatterplot.)

2. Obtain the least-squares regression line. Generally, you will be able to use a calculator or a computer to do this. Again, Chapter 2 is the source for the details on the line.

3. Make sure there are no outliers or influential observations. Both of these will influence the regression line and our inference methods.

4. Obtain the correlation coefficient, r, and square it to get r^2 (also in Chapter 2).

Recall that the equation of the regression line is $\hat{y} = a + bx$, where a is the intercept and b is the slope. Remember that $\hat{y}$ means "the predicted value of y." This regression line is for the sample data we collected. The **true regression line** is $\mu_y = \alpha + \beta x$. The true unknown slope is β, and μ_y is the mean value of y at the given values of x, and the true intercept is α; these are the population parameters. The statistics we will use are a and b, which are unbiased estimates of α and β.

KEY CONCEPTS

Assumptions for inference

- The relationship between x and y is linear. Check this by looking at the scatterplots of the data and the residuals. The data should show a linear trend but the residuals, plotted against x, should look random.

- The standard deviation is the same for all values of x. Again check both scatterplots. There should be no change in the spread of the points.

- Always check for normality. Use the residuals to make a histogram or stemplot and check it for skewness and other departures from normality. This is especially important for the prediction intervals.

Statistics and parameters

Parameters	Statistics that estimate them
α	a
β	b
σ (the standard deviation of y)	s

The standard error about the least-squares line is $s = \sqrt{\dfrac{1}{n-2}\sum(y - \hat{y})^2}$ The residuals (observed y – predicted y) are $(y - \hat{y})$ and they estimate how much y varies about the true line. The standard error has $n - 2$ degrees of freedom. Usually s will be calculated by a calculator or software, but if you have to do it yourself, follow these steps:

1. Find each predicted $\hat{y}$ for each x in the data.

2. Calculate the residuals, $(y - \hat{y})$.

3. Finally, use the formula to find s.

Inference for the slope

To calculate a **confidence interval** for the slope we need the estimate + the margin of error. For the slope, the estimate is b and the margin of error is

$$t * \left(\frac{s}{\sqrt{\sum (x - \bar{x})^2}} \right)$$

where t^* is the upper $(1 - C)/2$ critical value from the t distribution with $n - 2$ degrees of freedom and the standard error of b (SE_b) is the rest. Remember Σ means "sum up."

Hypothesis tests for the slope test the null hypothesis

$$H_0: \beta = 0$$

that there is no linear relationship between x and y. The test statistic is $t = \dfrac{b}{SE_b}$, the standardized value of b. Find the P-values for the appropriate alternative hypothesis, where T has the t distribution with $n-2$ degrees of freedom.

$$\text{If } H_a \text{ is } \beta > 0 \text{ find } P(T \geq t).$$
$$\text{If } H_a \text{ is } \beta < 0 \text{ find } P(T \leq t).$$
$$\text{If } H_a \text{ is } \beta \neq 0 \text{ find } 2P(T \geq t).$$

You may still find it helpful to draw a picture to see exactly what area you are looking for.

Inference for the predictions

We can make predictions easily just by plugging in the value of x into the formula to get the associated value of y. However, our prediction will carry much more weight if we can give a **confidence interval** for it. Actually, there are two kinds of predictions: predictions for a mean value of y and predictions for a specific value of y. The intervals have the same estimates, $\hat{y}$, but the margins of error are different. If you want to estimate a mean response, call the interval a confidence interval. If you want to estimate an individual response, call the interval a **prediction interval**.

For a prediction interval the margin of error is $t^* \, SE_{\hat{y}}$, where

$$SE_{\hat{y}} = s\sqrt{1 + \frac{1}{n} + \frac{(x^* - \bar{x})^2}{\sum(x - \bar{x})^2}}$$

and t^* is the upper $(1-C)/2$ critical value of the t distribution with $n-2$ degrees of freedom.

For a confidence interval the margin of error is $t^* \, SE_{\hat{\mu}}$, where

$$SE_{\hat{m}} = s\sqrt{\frac{1}{n} + \frac{(x^* - \bar{x})^2}{\sum(x - \bar{x})^2}}$$

In both settings t^* is the upper $(1-C)/2$ critical value of the t distribution with $n-2$ degrees of freedom and x^* is the particular value of x we are interested in.

ANSWERS TO SELECTED TEXT EXERCISES

Exercise 10.1

(a) If we look at the data, we see that as the length of the femur increases, so does the length of the humerus. Thus there is a positive association between femur and humerus length. A scatterplot of the data with femur length as the explanatory variable is given below.

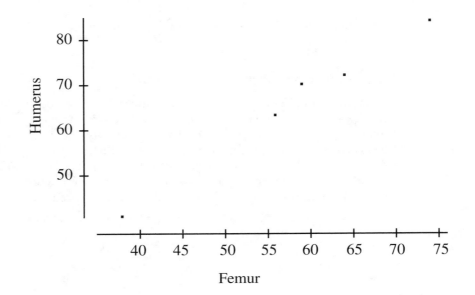

The scatterplot indicates a fairly strong positive association between femur and humerus length. If we calculate the correlation r and the equation of the least-squares line we obtain the following:

$$r = 0.994$$

$$\text{humerus length} = -3.65959 + 1.19690(\text{femur length})$$

The correlation is very high, so that one would expect that femur length would allow good prediction of humerus length.

(b) The slope β of the true regression line tells us the mean increase (in cm.) in the length of the humerus associated with a 1 cm. increase in the length of the femur in *Archaeopteryx*. Our estimate of β from the data is

$$\beta = 1.19690$$

the slope of the least-squares regression line. Our estimate of α from the data is

$$\alpha = -3.65959$$

the intercept of the least-squares regression line.

(c) The residuals for the five data points are given below

Observed value of humerus length	Predicted value of humerus length −3.65959 + 1.19690(femur length)	Residual (observed − predicted length)
41	41.822618	−0.822618
63	63.366820	−0.366820
70	66.957520	3.042480
72	72.942021	−0.942021
84	84.911022	−0.911022

The sum of the residuals listed is −.000001, the difference from 0 due to roundoff. To estimate the standard deviation σ in the regression model, we first calculate the sum of the squares of the residuals listed. We get

$$\sum \text{residual}^2 = 11.785306$$

Our estimate of the standard deviation σ in the regression model is therefore

$$s = \sqrt{\frac{1}{n-2}\sum \text{residual}^2} = \sqrt{\frac{1}{5-2}(11.785306)} = 1.982028$$

Exercise 10.5

(a) The equation of the least-squares regression line is, from the statistical software output (the values in the coeff column give the intercept and slope)

humerus length = −3.6596 + 1.1969(femur length)

(b) The value of the t statistic for testing H_0:$\beta = 0$ is determined by the equation $t = \frac{b}{SE_b}$, where b is the slope of the least-squares regression line and SE_b is the standard error of the least-squares slope. Both these quantities are given on the computer output, so

$$t = \frac{b}{SE_b} = \frac{1.1969}{0.0751} = 15.9374$$

(c) Recall that the sample size was $n = 5$. Thus t has $n - 2 = 5 - 2 = 3$ degrees of freedom. The P-value of t against the one-sided alternative H_a:$\beta > 0$ can be determined from the $df = 3$ row of Table C. The relevant portion of this table is

$df = 3$

p.	0005
t^*	12.92

and the P-value is thus < .0005.

Exercise 10.9

(a) Here is a scatterplot showing the relationship between power boats registered and manatees killed. Since we are trying to use the number of power boat registrations to explain the number of manatees killed, power boat registrations is the explanatory variable.

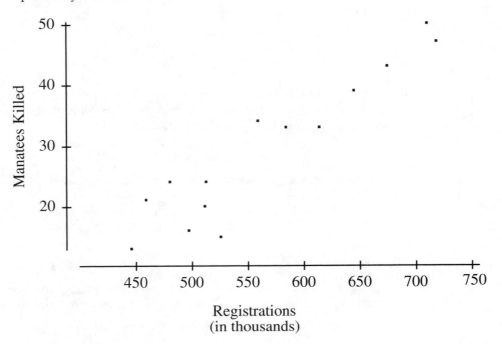

(b) The overall pattern is roughly linear with a positive slope. There are no clear outliers or strongly influential data points.

(c) We recall that r^2 tells us the percentage of the variation in the response variable accounted for by the explanatory variable in the least-squares regression line. In this case $r^2 = .886$, thus 88.6% of the variation in manatees killed is accounted for by power boat registrations.

(d) The slope β of the true regression line tells us the mean increase in the manatee deaths associated with an increase of 1000 power boat registrations. A 90% confidence interval for β is determined by the equation $b \pm t^*SE_b$, where $b = 0.12486$ is the slope of the least-squares regression line, t^* is the upper $(1 - C)/2 = (1 - .90)/2 = .05$ critical value from the t distribution with $n - 2 = 14 - 2 = 12$ degrees of freedom, and $SE_b = 0.01290$ (obtained for the Minitab output given in the problem) is the standard error of the least-squares slope b. From Table C we find $t^* = 1.7823$. Thus our 90% confidence interval is

$$b \pm t^*SE_b = 0.12486 \pm (1.7823)(0.01290) = 0.12486 \pm 0.02299 = (0.10187, 0.14785)$$

(e) Using the least-squares regression line, if Florida decided to freeze power boat registrations at 700,000 we would predict

$$\text{manatees killed} = -41.430 + 0.12486(\text{power boat registrations, in thousands})$$
$$= -41.430 + 0.12486(700)$$
$$= 45.972$$

(f) The prediction of 45.97 agrees with our answer in (e), after rounding. Since we want to estimate the *mean* number of manatees that would be killed each year if

Florida froze boat registrations at 700,00, we use a 95% confidence interval for the mean. The Minitab output given in the problem indicates that this 95% confidence interval is

$$(41.49, 50.46)$$

Exercise 10.13

(a) Here is the desired scatterplot.

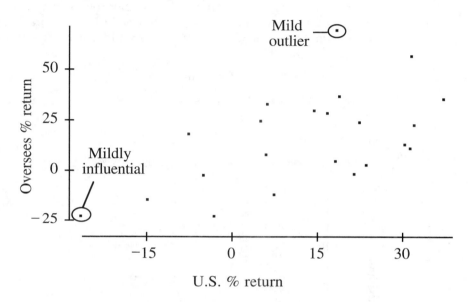

(b) The value of r^2 is given on the output as 32.4% or 0.324. Thus $r = \sqrt{0.324} = 0.569$. The relationship appears to be very roughly linear with a positive slope, but with considerable variation about a linear trend. The correlation of r = 0.569 suggests a moderate amount of positive association. The value of $r^2 = 0.324$ tells us that 32.4% of the variation in % return on overseas investments is accounted for by corresponding % return on U.S. investments. Again this is only a small to moderate percentage of the total variation in % return on overseas investments. This means that % return on U.S. investments will be only moderately valuable in predicting % return on overseas investments.

The scatterplot does not show any strong outliers or strongly influential observations. A mild outlier and mildly influential observation are circled in the plot.

(c) If we assume the mean % return on overseas investments, y, has a straight-line relation with % return on U.S. investments, x, then we are assuming a model of the form

$$\mu_y = \alpha + \beta x$$

To test for no linear relation, we test the hypotheses

$$H_0 : \beta = 0$$
$$H_a : \beta \neq 0$$

The Minitab output given in part (b) tells us that the test statistic (t-ratio for the predictor Overseas) is 3.09 with P-value 0.006. This indicates that there is strong evidence that a linear relationship is useful in describing the relation between % return on overseas investments and % return on U.S. investments.

(d) The least-squares regression line has (using the information in the Minitab output given in (b)) the equation

overseas % return = 4.777 + 0.8130(U.S. % return)

For a U.S. % return of 20%, we would predict

overseas % return = 4.777 + 0.8130(20%) = 21.037%

(e) The value we calculated in (d) agrees, after roundoff, with the "Fit" in the computer output. If we want to give a 95% interval for the return of foreign stocks *next* year if returns on U.S. stocks are 20%, we should use a 95% prediction interval. The output tells us that this interval is

(−21.97%, 64.04%)

Unfortunately, this interval is so wide as to be essentially useless in practice.

Exercise 10.15

(a) Here is a plot of the residuals against U.S. % return.

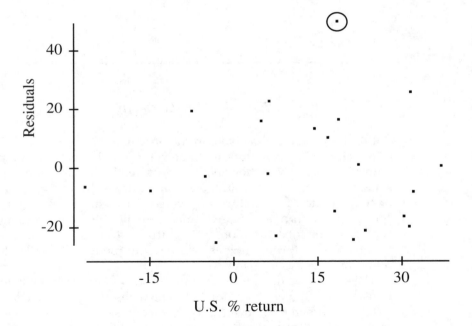

There is a mild outlier in the scatterplot (the point circled), which might suggest a violation of the assumption that the residuals have an approximate normal distribution.

(b) Here is a histogram of the residuals (you might also try making a stemplot).

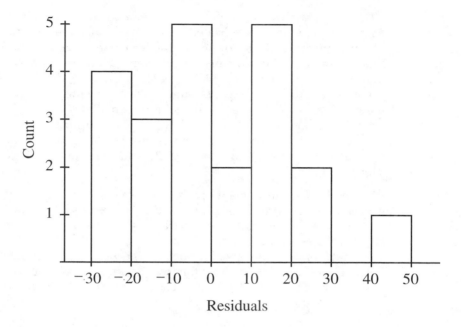

We see there is a possible outlier, the residual with value 49.501. We have circled this in the scatterplot in (a).

CASE STUDY

Smart people are often called "eggheads." Aliens always seem to be drawn with small bodies and large heads. Is the size of your brain an indicator of your mental capacity? The debate over the relationship between intelligence and brain size has been waged throughout most of written history. Early investigations in this area of research were crude, as researchers simply measured the size of a subject's head. Later studies included measurements of the size and weight of the brain taken after a subject had died. Many of the results of these early studies into the connection between brain size and intelligence have since been dismissed because of the crude manner in which the measurements were taken and/or experimental bias on the part of the researchers. Much of this early research was, unfortunately, used to perpetuate existing racial stereotypes. A fascinating discussion of these issues can be found in the book *The Mismeasure of Man* by Stephen J. Gould.

In spite of the biases present in early research, the debate over the relation between brain size and intelligence continues. Advancements in science and technology have led to many new ways to measure the "size" of the brain. In a recent study, L. Willerman, R. Schultz, J. N. Rutledge, and E. Bigler (1991), used Magnetic Resonance Imaging (MRI) to determine the brain size of their subjects. The researchers took into account gender and body size to draw conclusions about the connection between brain size and intelligence.

The researchers conducted their study at a large southwestern university. They describe the selection of their sample as follows:

The final sample consisted of 40 right-handed Anglo introductory psychology students who had indicated no history of alcoholism, unconsciousness, brain damage, epilepsy, or heart disease. These subjects were drawn from a larger pool of introductory psychology students with total Scholastic Aptitude Test Scores of ≥ 1350 or ≤ 940 who had agreed to satisfy a course requirement by allowing the administration of four subtests (vocabulary, similarities, block design, and picture completion) of the Wechsler (1981) Adult Intelligence Scale-Revised. With prior approval of the University's research review board, students selected for MRI were required to obtain prorated full-scale IQs of ≥ 130 or ≤ 103, and were equally divided by sex and IQ classification.

Magnetic Resonance Imaging (MRI) scans were performed at the same facility for all 40 subjects. Several images were taken of each subject's brain. A computer counted the number of relevant pixels (dark dots representing brain tissue) and the total count served as an index for brain size.

The data for the study are given below. The weights of two subjects and the height of one subject were withheld by the researchers for reasons of confidentiality. These missing values are indicated by a •. The particular variables listed are the following.

Gender = Male or female
FSIQ = Full Scale IQ scores based on the four Wechsler (1981) subtests
Weight = body weight in pounds
Height = height in inches
MRI count = total pixel count from 18 MRI scans

Gender	FSIQ	Weight	Height	MRI count
Female	133	118	64.5	816932
Male	140	•	72.5	1001121
Male	139	143	73.3	1038437
Male	133	172	68.8	965353
Female	137	147	65.0	951545
Female	99	146	69.0	928799
Female	138	138	64.5	991305
Female	92	175	66.0	854258
Male	89	134	66.3	904858
Male	133	172	68.8	955466
Female	132	118	64.5	833868
Male	141	151	70.0	1079549
Male	135	155	69.0	924059
Female	140	155	70.5	856472
Female	96	146	66.0	878897
Female	83	135	68.0	865363
Female	132	127	68.5	852244
Male	100	178	73.5	945088
Female	101	136	66.3	808020
Male	80	180	70.0	889083
Male	83	•	•	892420
Male	97	186	76.5	905940
Female	135	122	62.0	790619
Male	139	132	68.0	955003
Female	91	114	63.0	831772
Male	141	171	72.0	935494

Female	85	140	68.0	798612
Male	103	187	77.0	1062462
Female	77	106	63.0	793549
Female	130	159	66.5	866662
Female	133	127	62.5	857782
Male	144	191	67.0	949589
Male	103	192	75.5	997925
Male	90	181	69.0	879987
Female	83	143	66.5	834344
Female	133	153	66.5	948066
Male	140	144	70.5	949395
Female	88	139	64.5	893983
Male	81	148	74.0	930016
Male	89	179	75.5	935863

Lets look at the data. We begin with a scatterplot of FSIQ (the response variable and the measure of intelligence) versus MRI count (the explanatory variable and indicator of brain size). Here is the plot.

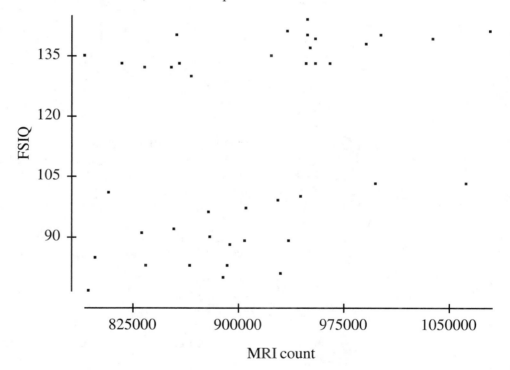

MRI count

The most striking feature of the plot are the two horizontal bands of points. Recall that only students with total Scholastic Aptitude Test scores of ≥ 1350 or ≤ 940 were eligible for the study. SAT scores are likely to be highly correlated with FSIQ scores, and the two horizontal bands of points reflect the two groups (those with SAT scores ≥ 1350 and those with scores ≤ 940).

Other than these two bands, nothing "leaps out" at us from the plot. Certainly the plot does not indicate any strong association between MRI count (brain size) and FSIQ score ("intelligence"). The sample correlation between these two variables is $r = 0.358$, indicating a weak positive association. Also $r^2 = 0.128$. This means that only 12.8% of the variation in FSIQ scores is accounted for by a straight-line relationship of FSIQ score with MRI count.

Many researchers argue that the association between brain size and intelligence is clearer if brain size is adjusted for "body size." Bigger people will have bigger heads and bigger brains, but this alone does not make them more intelligent. What is important is if their brain size is unusually large for their body size. There are many ways to attempt to adjust for body size. Both the weight and the height of a person can be used as a measure of body size. A simple adjustment is to divide brain size (MRI count) by weight or by height. The resulting ratio might be interpreted as brain size relative to body size (i.e., brain size per pound or per inch). From the data we create the ratios MRI count/weight and MRI count/height. Scatterplots of these ratios versus MRI count are given below.

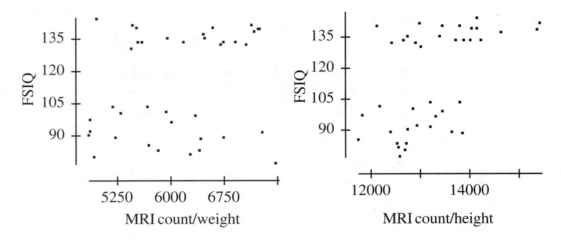

The correlation between FSIQ and MRI count/weight is $r = 0.235$ while the correlation between FSIQ and MRI count/height is $r = 0.518$. It would appear that using height as a measure of body size and comparing FSIQ to MRI count/height displays the association between "intelligence" and (relative) brain size more clearly. The scatterplots confirm that the association looks stronger when we use the ratio MRI count/height.

If we run a formal regression analysis on the straight line relation between FSIQ and MRI count/height using statistical software we obtain the following:

Dependent variable is FSIQ

R squared = 26.8%

$s = 20.71$

Variable	Coefficient	S.E. of Coeff	t-ratio	P-value
Constant	−76.2384	51.87	−1.47	0.1501
MRI count/height	0.014348	0.0039	3.68	0.0007

We notice that the P-value for the test that the slope of the least-squares regression line is 0 is very small. This is strong evidence that the slope is nonzero (and, in fact, positive). Recall this also means that the data provide strong evidence that the correlation between FSIQ to MRI count/height is nonzero (and, in fact, positive). Also note that the data on one subject are not used in this analysis since information concerning his height is missing (see the data).

Researchers also speculate that the relationship between "intelligence" and brain size may be different for men and women. To investigate this, we make a scatterplot of FSIQ scores versus our new ratio MRI count/height, and use the plotting symbol o to represent women and the plotting symbol x to represent men. Here is the plot.

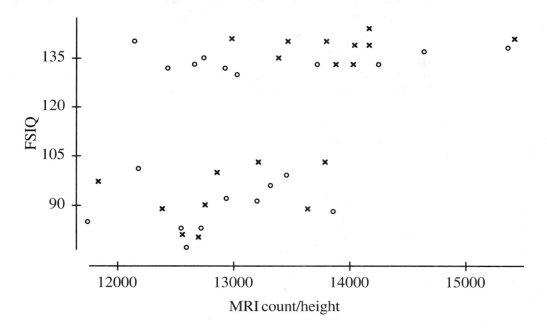

The correlation between FSIQ and MRI count/height for women can be calculated (you can do it yourself if you like) and is $r = 0.347$ while for men $r = 0.692$. The stronger correlation for men can be seen by observing that in the upper horizontal band of points, the men (indicated by the symbol x) tend to lie to the right of the women. The data do suggest that the strength of the association depends on gender.

Do these data provide evidence that a bigger brain means you are smarter? As in all observational studies, we must be clear about what the data actually show. First, remember association does not imply causation. This is an observational study, not a designed experiment. Second, it is not clear what population is represented by the student volunteers in the study. This was not a random sample from some population. Only right-handed Anglo introductory psychology students who had indicated no history of alcoholism, unconsciousness, brain damage, epilepsy, or heart disease and with SAT scores ≥ 1350 or ≤ 940 were even eligible to take part. Third, the strength of the observed correlation between FSIQ and MRI count/height must be interpreted with caution. We did multiple analyses and chose the measure of "body size" that produced the highest correlation between FSIQ and MRI count/"body size." This sort of approach yields results that inflate or overstate the strength of the evidence in the data. Even so, a correlation of 0.518 has $r^2 = 0.268$, so that only 26.8% of the variation in FSIQ scores is accounted for by a straight-line relationship of FSIQ score with MRI count/height. The association is not overwhelming. Fourth, the researchers used a very select pool of students and only students with either high or low SAT scores and hence only high or low FSIQ scores. While such a study design is useful for detecting an association, it does tend to inflate the strength of the association. Finally, we must be careful not to confuse FSIQ scores with "intelligence." The book *The Mismeasure of Man* by Stephen J. Gould discusses this issue in some detail.

Does all this mean the results are worthless? Not entirely. No designed experiment is possible (how would one assign "brain size"?). Like the case for the association between lung cancer and smoking, this study might be taken as one piece of evidence in a larger case. We would interpret the results of this study as intriguing evidence of a weak positive association between FSIQ scores and MRI count/height. We would want to see more evidence, however, before we felt the evidence for an association was convincing.

Reference
Willerman, L., Schultz, R., Rutledge, J. N., and Bigler, E. (1991), "In Vivo Brain Size and Intelligence," *Intelligence*, **15**, 223–228.

GLOSSARY

The page number that follows each definition refers to the textbook page on which the term first appears.

acceptance sampling   When a sample is taken on the decision to either accept or reject the population based on the sample. Often used in manufacturing situations. (p. 387)

α Alpha, the Greek letter representing the significance level. (p. 361)

alternative hypothesis, H_a   The hypothesis that states what we believe is true, that an effect is present. What we are trying to show using the data. (p. 351)

analysis of variance F test   Overall test of the hypothesis that all populations have the same mean. (p. 560)

ANOVA   Abbreviation of analysis of variance. (p. 560)

ANOVA table   Table listing the sum of squares, mean squares, and degrees of freedom for an analysis of variance F test. (p. 578)

association, negative   When above average values of one variable are associated with below average values of another variable. (p. 99)

association, positive   When above average values of one variable are associated with above average values of another variable and below average values are associated with below average values. (p. 99)

bar chart   A graph that lists the values of a categorical variable on a horizontal scale and places a bar, whose height represents the number of times that value occurs in the data, above each value. (p. 13)

bias   The systematic favoring of a certain outcome. (p. 181)

binomial coefficient $\dbinom{n}{k} = \dfrac{n!}{k!(n-k)!}$. (p. 284)

binomial distribution   The distribution of the count of X successes in n observations when the probability of a success in any trial is p. Must meet the rules of the binomial setting. X is a discrete random variable taking values from 0 to n. (p. 281)

block   A group of experimental units that are similar in a way that may effect the results of the experiment. (p. 214)

block design   When the random assignment of units to treatment is done separately within each of the blocks of similar units. (p. 214)

boxplot   A picture form of the five-number summary. (p. 44)

C The confidence level. (p. 332)

categorical variable   A variable that records to what group or category an individual belongs. For example, hair color is a categorical variable. (p.12)

center   The "middle" of the data. (p. 18)

central limit theorem   When the sample is an SRS, n is large, and the population has mean μ, and finite standard deviation σ, then the distribution of the sample mean will be close to normal, $N(\mu, \sigma/\sqrt{n})$, regardless of the shape of the population distribution. (p. 298)

chi-square distribution   A family of distributions that take only positive values and are skewed to the right. The distribution associated with the chi-square statistic in testing the hypothesis of no relation in a two-way table. (p. 533)

chi-square statistic, X^2   A measure of how far the observed counts in a two-way table are from the expected counts. (p. 528)

chi-square test   The test of the hypothesis of no relation in a two-way table. (p. 533)

column total   The sum of all the individual observations that are in a column of a two-way table. (p. 151)

comparison   When two or more groups are looked at relative to each other. Is one better than the others? (p. 203)

completely randomized   When all the experimental units in an experiment are assigned at random to all the treatments. (p. 206)

conditional distribution   The distribution of a particular row or column in a two-way table. More generally, the distribution of some variable only for those cases satisfying some condition. (p. 155)

confidence interval   A range of values in which a parameter is believed to lie within a specified degree of certainty. More specifically, an interval computed from sample data by a method that has a pre specified probability of producing an interval containing the true value of the parameter. (p. 329)

confidence level   The probability statement associated with a confidence interval. The probability that the method we use to produce a confidence interval will yield an interval containing the true value of the parameter. (p. 332)

confounding   When the effects of two or more variables are mixed together and you cannot tell the effects apart. (p. 179)

continuous random variable   A random variable that can take an infinite number of values. (p. 262)

control chart   Statistical tools used to monitor a process and alert the user when the process changes. (p. 305)

control group   The subjects in an experiment that receive no actual treatments. Often control groups are given a placebo rather than an actual treatment. (p. 203)

control limits   The upper and lower lines on a control chart. Points beyond the control limits signal a change in the process. (p. 308)

convenience sampling   A very poor method of sampling that chooses the individuals who are most accessible. (p. 180)

correlation   A measure of the linear association between two quantitative variables. Correlations range from –1 (the data lie exactly on a negatively sloping straight line) to +1 (the data lie exactly on a positively sloping straight line). (p. 111)

counts   How many in each category. (p. 276)

critical value   The number from a distribution used to calculate a confidence interval or a significance test. The cutoff point in a significance test. A value of a distribution having a specific upper tail probability. (p. 334)

data   A collection of information, often numerical, about a group of objects. (p. 10)

degrees of freedom   The number of observations, n, minus 1. It is used when calculating the variance and standard deviation. More generally, it represents the number of 'unrelated' numbers used in calculating a quantity. (p. 49)

density curve   A smooth curve covering an area equal to one. Density curves are idealized descriptions of the distribution of data. (p. 56)

deviation   The difference between a value and the mean. (p. 46)

discrete random variable   A random variable that can take only a finite number of values. (p. 252)

distribution   What values a variable takes and how often it takes each one. (p. 12)

double-blind experiment   When neither the subjects nor the experimenters having contact with subjects know which treatment a subject received. (p. 214)

experiment   A controlled study where the explanatory variable is manipulated and the response variable is observed. (p. 147)

experimental units   The individuals on which the experiment is done. (p. 199)

explanatory variable   A variable that is used to explain the response variable. (p. 94)

exploratory data analysis   Using graphical and numerical summaries to examine data for the purpose of uncovering patterns, trends, and other interesting features. (p. 12)

exponential distribution   A special distribution shape that can sometimes be used to describe the lifetime of objects or the waiting time until some event. (p. 298)

extrapolation   Using the least-squares regression line to make a prediction based on a value of the explanatory variable, x, that is outside the range of the values used to create the regression line. These predictions are often unreliable. (p 142)

F distribution   The family of distributions of the F statistic, defined by a pair of degrees of freedom. (p. 464)

F statistic   The statistic used to compare the standard deviations of two populations. Use with caution for comparing two standard deviations since this procedure is not robust. (p. 465)

$F(j,k)$ Shorthand for the F statistic where j is the numerator's degrees of freedom and k is the denominator's degrees of freedom. (p. 464)

factorial $n! = n \times (n-1) \times (n-2)\dots 3 \times 2 \times 1$. (p. 284)

factors   The explanatory variables in an experiment. (p. 199)

five-number summary   A description of a data set consisting of the minimum, first quartile, median, third quartile, and maximum. (p. 44)

H_0   The null hypothesis. (p. 351)

H_a   The alternative hypothesis. (p. 351)

histogram   A graphical display of a quantitative variable. It uses bars of equal width to show how often a value or set of values occur. (p. 15)

in control/out of control   When the variable being monitored has shifted and is no longer the same as it once was. (p. 308)

independent   When knowing the result of one event does not give you any additional information about the probability of another event. (p. 280)

individuals   The objects (people, animals, plants, etc.) described by the data and about which more information is desired. (p. 10)

inference   Drawing conclusions about a population from information in a sample from the population. (p. 324)

influential observation   An observation that if removed will noticeably change the position of the least-squares regression line. (p. 135)

intercept   Where a line crosses the vertical, or y, axis. The value of $\hat{y}$ when $x = 0$. In least-squares regression the intercept is $a = \bar{y} - b\,\bar{x}$. (p. 123)

law of large numbers   Describes how the mean of many observations of a random variable will get closer and closer to the true mean of the variable. (p. 259)

least-squares regression line   A regression line that minimizes the sum of the vertical distances of the points to the line. (p. 121)

level   The different settings or values of a factor. (p. 199)

linear relationship   When the relationship between two variables can be described well by a straight line. (p. 101)

lurking variable   A third, hidden, variable that may be influencing the studied variables. For example, the number of crimes committed by teenagers increases as the number of hours of daylight increases. This positive association is deceptive as there are more hours of sunlight in the summer and thus more free time and less adult supervision. (p. 143)

margin of error   The "give or take" (±) amount in a confidence interval. How accurate we believe our estimate of a parameter is. (p. 329)

marginal distribution   The distribution of the row variable or the column variable in a two-way table. (p. 151)

matched pairs design   A special type of block design that uses only two units in each block and the units are as similar as possible. The units are assigned to treatments at random. (p. 215)

matching   A process where each subject or experimental unit in one treatment group is paired with a subject or unit in another treatment group that is as similar to them as possible. (p. 204)

mean   A measure of center, the numerical average, represented as $\bar{x}$. The mean has an interpretation as the center of mass if the data are represented as masses on a horizontal axis. (p. 36)

mean of a density curve   The balance point of a density curve, if the curve were a solid and resting on a see-saw. (p. 58)

mean square for error, (MSE)   The ratio of the sum of the variances times their degrees of freedom for each treatment group or population to the degrees of freedom. (p. 577)

mean square for groups, (MSG)   The mean square for the averages of each treatment group or population. (p. 577)

mean squares   The ratio of the sum of the squared deviations from the overall mean to the degrees of freedom. (p. 576)

median   A measure of center where half of the observations are less and half are greater than the median, the middle number . (p. 38)

median of a density curve   The point on a density curve where the area on either side is equal. (p. 58)

MSE   Abbreviation of the mean square for error. (p. 577)

MSG   Abbreviation of the mean square for groups. (p. 577)

μ   The Greek letter mu. An abbreviation for the mean of a density curve. (p. 59)

multistage sample   A complex sampling method in which individuals are selected in a series of stages. (p. 188)

N(μ, σ)   A shorthand way of writing the phrase 'normal distribution with mean = μ and standard deviation = σ. (p. 63)

nonresponse   When individuals who are selected do not participate or cannot be contacted. (p. 190)

normal distribution   A class of density curves that all have a symmetric, bell-shaped curve. Each has the same general shape and is defined by its mean, μ, and its standard deviation, σ. (p. 60)

null hypothesis, H_0   The hypothesis that states there is no effect, only chance is at work. (p. 350)

observational study   When data are gathered by observing subjects with no active interventions by the researcher. (p. 198)

one-sided alternative   When the alternative hypothesis indicates an effect (deviation from the null hypothesis) in one direction only, either greater than or less than. (p. 357)

outlier   An observation that stands out from the rest for some reason. (p. 18)

P(A)   Probability of the event *A* occurring. (p. 255)

$\hat{p}$ Called "*p* hat". It is the sample proportion, a statistic used to estimate *p*, the population proportion. (p. 232)

$\hat{p}$ chart   A control chart that monitors a process by tracking the sample proportion, $\hat{p}$, over time. (p. 310)

P-value   The observed significance level. It is the chance of getting a result as extreme or more extreme than the one observed. Small *P*-values support the alternative hypothesis. (p. 353)

parameter   A number describing some aspect of the population we are interested in. (p. 230)

pie chart   A circle that is divided into slices (like a pie) each of which corresponds to a value of a categorical variable. The size of a slice (proportion of the total area of the circle) corresponds to the proportion of time that value occurs in our data. (p. 13)

placebo effect   When the results of an experiment are not due to the specific treatment but due to the fact that some treatment, even a fake, was administered. (p. 202)

pooled sample proportion   When the results of several samples are combined together to produce the proportion, which is the total number of successes over the total number of observations. (p. 505)

pooled standard deviation   Square root to the MSE. (p. 579)

population   The whole group you are interested in learning something about. (p. 180)

population proportion   The proportion of how many in the population belong to a certain category. (p. 268)

prediction   Using what you know to make an educated guess at what you don't know. In regression, the regression line can be used to predict a value of *y* given a certain value of *x*. (p. 120)

prediction interval   Like a confidence interval but for estimating (predicting) the next response. (p. 607)

probability   The long run proportion of times some event occurs. (p. 243)

probability distribution   The distribution of a random variable, tells us the values the variable takes and the likelihood of each value. (p. 251)

probability histogram   A histogram of a discrete random variable, the heights of the bars represent the probability of that value occurring. (p. 254)

probability sample   A collection of individuals selected from the population in a way so that each individual has a known and non-zero chance of being selected. (p. 187)

population mean   The mean of all the values in a population, denoted μ. (p. 297)

quantitative variable   A variable that has numerical values and with which it makes sense to do arithmetic. It is a quantity. For example, the number of magazines subscribed to is quantitative. We could ask for an average number of subscriptions or a maximum. (p. 12)

quartiles   Produced by sorting the data in increasing order and splitting the data into quarters. (p. 42)

r The correlation coefficient. (p. 111)

r^2 The square of the correlation coefficient, r. Shows the goodness of the fit of the regression line. In particular, it gives the proportion of the variation in the response variable that is explained by the least-squares regression line. (p. 126)

randomization   Using a chance mechanism such as a table of random digits to make the assignments of subjects to treatments, level to treatment, or subjects to experimental groups. (p. 205)

random variable   A variable whose value is the outcome of a random event. (p. 250)

range   The difference between the highest and lowest observed values. The range shows the full extent of the data. (p. 41)

regression line   A straight line that describes how a response variable changes as the explanatory variable changes. (p. 119)

replication   Repetition. Repeating an experiment on many subjects or units. (p. 209)

residual   The error in a prediction. Residual = $y - \hat{y}$. (p. 129)

residual plot   A scatterplot of the residuals against the explanatory variable or some other variable. (p. 131)

response bias   When individuals who participate in a survey do not respond truthfully due to the way the question is worded, the presence of an observer, fear of a negative reaction from the interviewer, or any other such source. (p. 191)

response variable   The variable that measures the outcome of a study. (p. 94)

robust   A statistical procedure that performs well even when the assumptions of the procedure are violated. (p. 424)

roundoff error   Errors in totals that occur due to rounding the individual numbers used to compute the total up or down. (p. 151)

row total   The sum of all the individual observations that are in a row of a two-way table. (p. 151)

sample   The part of the whole that you use to answer questions about the whole or use in an experiment. (p. 180)

sample design   The plan of an experiment, survey, study, or sample. (p. 180)

sample mean   The mean of all the values in a sample, denoted $\bar{x}$. (p. 297)

sample proportion   The proportion of how many in a sample belong to a certain category. This is used to estimate how many in the population belong to the category. (p. 268)

sampling   Choosing part of a population for the purpose of gaining information about the entire population. (p. 178)

sampling distribution   The distribution of the values of a statistic in all possible samples of the same size from the same population. (p. 234)

sampling frame   The list of individuals from which a sample is selected. (p. 192)

sampling variability   The variation in the value of a statistic that we would see if we took many samples under the same conditions and measured the statistic again and again. (p. 232)

scatterplot   A graph that shows the values of two quantitative variables per observation, one (the explanatory variable if there is one) on the horizontal axis and one (the response variable, if there is one) on the vertical axis. (p. 97)

shape   What the distribution looks like. For example, symmetric, skewed, bell-shaped, flat, and rectangular are some words typically used to describe the shape of a distribution. (p. 18)

σ The Greek letter sigma. An abbreviation for the standard deviation of a density curve. The variance of a density curve is abbreviated σ^2. (p. 59)

significance level   A fixed probability that the P-value must be smaller than in order to declare the data statistically significant. How unlikely the data must be under the null hypothesis before we are willing to attribute the results to an effect other than chance. (p. 361)

significance test   A method of inference for choosing between specific hypotheses. Can observed results be plausibly explained by chance or is there a real effect present? (p. 350)

simple random sample   A collection of individuals from the population chosen in a manner so that each possible set of individuals has an equal chance of being selected. (p. 183)

Simpson's paradox   The reversal of the direction of a comparison or association when data from several groups are combined to form a single group. (p. 159)

simulation   An imitation. An artificial representation of a real event or process used to study the real event or process. (p. 232)

skewed   The opposite of symmetric. The observations are clustered to one side of the center and spread far out along the other side (this latter side being the direction of the skewness). (p. 20)

slope   The amount y changes when x increases by 1. In least-squares regression, the slope is $b = r\dfrac{s_y}{s_x}$. (p. 123)

spread   The range of the data; the extent of the data; how far the values are from each other. (p. 18)

SRS   Simple random sample. (p. 183)

standard deviation   A measure of spread based on the mean that tells how far the observations are from the mean on average. σ is its abbreviation. (p. 47)

standard error   The estimate of the standard deviation of a statistic. (p. 409)

standard normal distribution   A specific normal distribution with mean 0 and standard deviation 1. Any normal distribution can be standardized and become standard normal. Abbreviated as $N(0,1)$. (p. 65)

standard normal table   A table of areas under the standard normal curve. For any z-score the table gives the area, or proportion, to the left of z under the curve. The table is Table A in the back and the inside front cover of the textbook. (p. 67)

standardizing   Creating a standard normal value or z-score for an observation. (p. 64)

statistic   A number we calculate based on a sample of the population without using any unknown parameters. (p. 231)

statistical control   When a variable continues to have the same distribution as it is observed over time. (p. 305)

statistical inference   The statistical methods that allow the user to answer specific questions from data with some guarantee that the answers are good ones. (p. 179)

statistical process control   Using control charts to monitor a process and determine if the process is in or out of control. (p. 311)

statistical significance   When an effect is too large to be reasonably attributed to chance. (p. 209)

stemplot   A graphical tool, easily constructed by hand and best used with small data sets for showing the shape of the data. A stemplot involves a stem (usually the left-most digits in the data) and leaves (the right-most digits in the data). (p. 23)

strata   (Singular: stratum) a layer or level. A group of individuals that has some characteristic or set of characteristics in common. (p. 187)

stratified sample   A sampling method where the population is divided into groups, called strata, of similar individuals on the basis of some common known characteristic and then an SRS is selected from each stratum. (p. 187)

subjects   Human experimental units. (p. 199)

success   In a two outcome situation, the outcome we are interested in. (p. 485)

symmetric   A distribution that looks the same on both sides of the center. (p. 20)

systematic random sample   A sampling method that is often used as part of a multistage sample, where a beginning point is chosen at random and then every so many individuals after that are selected for the sample. This is not an SRS. (p. 196)

t distribution   The family of distributions of the t statistic, each member defined by its degrees of freedom. (p. 409)

$t(k)$ Shorthand for the t distribution with k degrees of freedom. (p. 410)

tests of significance   *See* significance test.

test statistic   A number used to measure the difference between what we observe in sample data and what we would expect to see if the null hypothesis were true. (p. 359)

time plot   A graph showing the values of a variable taken over time. Time is commonly on the horizontal axis. (p. 26)

treatment   The combination of explanatory variables applied to a particular unit. (p. 199)

trends   A pattern in a plot. Often, but not always, a trend refers to a pattern over time. For example, the average monthly temperature measured in degrees Celsius will show a trend that repeats each year. (p. 28)

two-sided alternative   When the alternative hypothesis says the effect (deviation from the null hypothesis) is any difference at all (i.e., the "not equal to" hypothesis). (p. 357)

two-way table   A table containing the counts of how many individuals fall into each pair of categories of two categorical variables. (p. 150)

Type I error   When H_0 is rejected but it is true. (p. 388)

Type II error   When H_0 is accepted but is false (H_a is true). (p. 388)

unbiased statistic   When the mean of the sampling distribution of a statistic is equal to the value of the population parameter the statistic is used to estimate. (p. 238)

undercoverage   When some group in the population is not given the chance to be in the survey. (p. 190)

units   A standard of measurement. For example, meters, pounds, and miles per hour are units. (p. 10)

variable   Any characteristic of an individual. Variables can have different values for different individuals. For example, the number of magazines a person subscribes to is a variable. Favorite color can also be a variable. (p. 10)

variance   The square of the standard deviation. σ^2 is its abbreviation. (p. 46)

voluntary response sample   When the people who participate in a study or survey select themselves by volunteering. Likely to produce much bias. (p. 178)

wording effects   the impact that the way a question in a survey or study is phrased can have on the way the question is answered. For example, a question that leads you to believe a certain answer is desired over another. (p. 191)

$\overline{x}$ The mean, the numerical average. (p. 36)

$\overline{x}$ **chart**   A control chart that monitors a process by tracking, the sample mean, $\overline{x}$, over time. (p. 309)

X^2 The symbol for the chi-square statistic. (p. 528)

$\hat{y}$ The predicted value of y. Called "y hat". (p. 122)

z^* The symbol for the critical value from the standard normal table. (p. 334)

z-score   The standardized value of an observation. (p. 64)